Technische Universität München
Lehrstuhl für Numerische Mathematik

Numerical Simulations in Room Acoustics using direct Coupling Techniques and Finite Elements

Martina Pospiech

Vollständiger Abdruck der von der Fakultät für Mathematik der Technischen Universität München zur Erlangung des akademischen Grades eines

Doktors der Naturwissenschaften (Dr. rer. nat.)

genehmigten Dissertation.

Vorsitzende: Univ.-Prof. Dr. Christina Kuttler
Prüfer der Dissertation: 1. Univ.-Prof. Dr. Peter Rentrop
                        2. Univ.-Prof. Dr. Gerhard H. Müller
                        3. Univ.-Prof. Dr. Rüdiger Weiner,
                           Martin-Luther-Universität Halle-Wittenberg

Die Dissertation wurde am 17.11.2010 bei der Technischen Universität München eingereicht und durch die Fakultät für Mathematik am 22.03.2011 angenommen.

Bibliografische Information der Deutschen Nationalbibliothek

Die Deutsche Nationalbibliothek verzeichnet diese Publikation in der
Deutschen Nationalbibliografie; detaillierte bibliografische Daten sind
im Internet über http://dnb.d-nb.de abrufbar.

ISBN 978-3-8325-3139-3

Logos Verlag Berlin GmbH
Comeniushof, Gubener Str. 47,
10243 Berlin
Tel.: +49 (0)30 42 85 10 90
Fax: +49 (0)30 42 85 10 92
INTERNET: http://www.logos-verlag.de

Contents

3

Acknowledgments

I would like to thank my supervisor Prof. Peter Rentrop for his support and encouragement during the past three years. I feel very privileged to have been working with him.

Further on, I would like to thank Prof. Gerhard Müller, Martin Buchschmid and Jovana Sremčević for their fruitful discussions. I always enjoyed the research cooperation and the social events with the Chair of Structural Mechanics. I remember with pleasure and with scientific passion the many days as Martin and I worked together sunken in difficult challenges literally spoken day and night.

In addition, I acknowledge Prof. Rüdiger Weiner for his help as the third supervisor and his helpful comments for this thesis.

I have to thank for the financial support by the International Graduate School of Science and Engineering (IGSSE). I am grateful for their given opportunity of the research exchanges to the Institute of Sound and Vibration Research (ISVR) in Southampton.

I thank to my supervisor Neil Ferguson, who spent a lot of time to talk with me about details in acoustics during my time at the ISVR. Moreover, the international colleagues of the Dynamic Group introduced me to various research fields and offered me a warm welcome in their group.

All above I thank to all staff and colleagues, especially to those at the Chair of Numerical Analysis. They were always open for scientific discussions and inspiring conversations and helped me sail through smoothly during the course of my present research.

I like to thank my friends for accompanying me all along my way in difficult times as well as in good times.

Last but by no means least, I like to thank my family for the opportunity to start and pursue a career in science. I am particularly indebted to my parents for their never-ending encouragement and ongoing support.

Abbreviations

BEM	Boundary Element Method
CAA	Computational Aeroacoustic
CFD	Computational Fluid Dynamic
CMS	Component Mode Synthesis
DFT	discrete Fourier transform
DTFT	discrete-time Fourier transform
dof	degree of freedom
FE	Finite Element
FFT	fast Fourier transform
FSI	fluid-structure interaction
FT	Fourier transform
frf	frequency response function
IDFT	inverse discrete Fourier transform
IDTFT	inverse discrete-time Fourier transform
IFT	inverse Fourier transform
Im	imaginary part of a complex number
MC	Monte Carlo
pdf	probability density function

Re	real part of a complex number
SEA	Statistical Energy Analysis
SEM	Spectral Element Method
SM	Spectral Method
SPL	sound pressure level
TPM	Theory of Porous Media
WBM	Wave Based Method

Scripts and Symbols

Subscripts

0	reference value/mean value
B	boundary at the interface
BC	related to the boundary condition
F	fluid (air)
inc	incident wave
$Load$	related to the load (sound source)
ref	reflected wave
TPM	related to porous structures

Superscripts

C	coupling mode
fi	fixed-interface normal modes
fr	free-interface normal modes
G	gas phase
H	homogeneous material
N	normal mode
nc	non-conservative
S	solid phase
T	transposed vector quality
$*$	linear fluctuation value of the related variable
$\widehat{}$	Fourier coefficient

Symbols

Latin Letters

A	area of sound source	$[m^2]$
C	damping coefficient (point impedance)	$[\frac{Ns}{m}]$
c	speed of sound	$[\frac{m}{s}]$
d	thickness of the absorber layer	$[m]$
e	sound energy density	$[\frac{J}{m^3}]$
$\boldsymbol{f}(\boldsymbol{x}, t)$	force in the fluid	$[\frac{N}{m^3}]$
f	frequency	$[Hz]$
I	acoustic intensity	$[\frac{W}{m^2}]$
i	imaginary unit	
K	stiffness of the boundary (point impedance)	$[\frac{N}{m}]$
k	wavenumber	$[\frac{1}{m}]$
$\boldsymbol{k}$	wave vector	$[\frac{1}{m}]$
L	Lagrangian	
M	mass of the boundary (point impedance)	$[kg]$
m	mass	$[kg]$
m_V	mass per volume by external sources	$[\frac{kg}{m^3}]$
n	volume fraction	
$\boldsymbol{n}$	normal outward vector	
P	sound power	$[W]$
$p(\boldsymbol{x}, t)$	total static pressure in the air	$[\frac{N}{m^2}]$

q	rate of volume velocity	$[\frac{m^3}{s}]$
r	reflection factor	
S	surface area	$[m^2]$
T	kinetic energy of the fluid	$[J]$
t	time	$[s]$
U	potential energy of the fluid	$[J]$
$U(\cdot, \cdot)$	uniform distribution	
V	volume of the geometry	$[m^3]$
$\boldsymbol{v}(\boldsymbol{x}, t)$	(total) sound velocity in the air	$[\frac{m}{s}]$
δW	virtual work	$[Nm]$
w	quadrature weight	
x	spatial coordinate in one dimension	$[m]$
$\boldsymbol{x}$	vector of spatial coordinates	$[m]$
Z	impedance	$[\frac{Ns}{m^3}]$

Greek Letters

α	absorption efficiency	
δ	Dirac delta function	
γ	heat capacity	$[\frac{J}{K}]$
λ	first Lamé constant	$[\frac{N}{m^2}]$
μ	second Lamé constant (shear modulus)	$[\frac{N}{m^2}]$
ν	test functions	
ξ	quadrature points	
$\rho(\boldsymbol{x}, t)$	total density of the air	$[\frac{kg}{m^3}]$
σ	standard deviation	
τ	transmission ratio	
Φ	velocity potential	
ψ	basis functions	
Ω	frequency	$[Hz]$
ω	circular frequency	$[\frac{rad}{s}]$

Contents

1 Introduction

1.1 Motivation

The understanding of the physical behaviour of sound in volumes of air enclosed by solid boundaries is vitally important for noise control and acoustic design. Exemplarily the acoustic comfort of passengers in road, rail and air vehicles is determined by mechanism of excitation, the geometry of the cavity and the material of its surrounding.

Since years the analysis of sound fields in closed cavities is of interest especially within the simulation of realistic boundary conditions. Absorbers and reflectors represent these realistic boundary conditions to optimize the sound field of the cavity dependent on the usage.

One gain of the optimization is to prevent disturbing effects of frequencies of excitation. Another gain is to optimize special characteristics of the sound field, amongst others the reverberation time and the spatial distribution of the sound field is the decay time of the sound pressure to a certain level after the sound source. For bigger rooms these characteristics are achieved by placing absorbers at the boundaries to absorb the sound energy. Interference between repeated reflections of sound waves from the boundaries of an enclosure that is not highly absorbent creates a spatially and temporally complex sound field. This sound field exhibits a frequency dependent behavior due to acoustic eigenmodes. The average distance between adjacent resonance frequencies decreases as the square of the frequency. Consequently it is difficult to impossible to identify individual resonances and related eigenmodes generally above about five times the fundamental resonance frequency. Because sound waves are repeatedly reflected, even very small variations from perfect regularity and uniformity of boundaries are sufficient to make enclosed sound fields unpredictable at any location inside the cavity. In addition, the uncertainty is increasing with frequency.

In spite of concerning oneself with affecting sound fields in complex shaped enclosed cavities, it is useful to analyse the acoustic behaviour of sound in geometrically regular cavities. The features of these cavities are generic to all enclosed fields and thus can provide a basis of qualitative understanding and enable the development of alternative models.

1

1.2 State of the Art

Computational Fluid Dynamics (CFD) methods are the basis of the discipline computational aeroacoustics (CAA), dated back to the middle of the 1980s, which analyses the sound generated by velocity and vorticity fields. The theory of room acoustics is an important branch in Technical Acoustics [79].

The main numerical approaches to describe the propagation in enclosed cavities for the low-frequency range are types of the Finite Element (FE) method and the Boundary Element Method (BEM) [75]. Both approaches can cope with an arbitrary geometry, but, for computational and accuracy reasons, are applicable primarily for low frequency predictions. These methods discretize the physical system and equations of motion and the modal degree of freedom (dof) are obtained. There are a couple of review papers on FEM and BEM [45, 46, 105].

Recently, a technique, called the Wave Based Method (WBM) [47], has been developed for the analysis of coupled vibro-acoustic problems in the mid-frequency range. It is based on the expansion of dynamic response variables in terms of wave functions, which are the exact solutions of the governing differential equations. These individual solutions are combined and constrained to solve the system with more general geometry or boundary conditions.

A common approach for the high-frequency range is the Statistical Energy Analysis (SEA) [58, 60, 70]. It is widely used by industry to predict noise caused by structural vibration and to optimize structural design for the minimization of vibration and noise. The SEA is robust for subsystems, if the averaging over frequency bands, points of excitation and points of observation is accomplished. However, its performance is limited if the spatial resolution of the response has to be described and if boundary conditions are investigated more in detail.

Finally, fluid-structure interaction (FSI) problems come into account, because of the fact that realistic complex boundaries have to be considered. FSI problems are related to numerical methods and present specific formulations in detail, see for instance [7, 18, 50, 81, 85]. Methods handling such problems are subdivided into two classes. The first one couples the fluid and the structure in an indirect way by exchanging the boundary conditions iteratively. The second class solves the whole system directly.

FE models including absorptive boundary conditions in acoustical computations have the drawback to lead to a huge number of dofs. In [62] an impedance approach considering a plane wave of incidence is proposed in order to reduce this number of unknowns . In [16] and in the scope of this thesis a wavenumber- and frequency-dependent impedance is used for plate-like compound absorbers to introduce varying angles of inclination for the sound waves [16, 17]. As a part of the compound absorber the porous foam is modelled by the Theory of Porous Media (TPM) [29]. An efficient way to reduce the computational effort for the optimization of the position of these acoustic elements or for averaging techniques is to simulate the coupled system in the frequency domain by applying a Component Mode Synthesis (CMS), which is a model order reduction method [96].

1.3 Layout of the Thesis

The underlying thesis presents a modal-based coupling method for linear room acoustic problems in the frequency domain. Therein we give an approach to approximate the sound fields of closed systems with different boundaries by using modal functions. These modal functions are the basis functions represented by normal and coupling modes. The modes are based on the velocity potential computed by an FE method. Related to the demanded sound field we consider time harmonic sound sources in the low-frequency range.

At the beginning of Chapter 2 we provide the basic theory to derive the governing equations for the coupling method. These equations are dependent on the physical quantities pressure and velocity and are given in the state formulation as well as in the frequency formulation. In addition all types of possible boundary conditions are presented for both formulations. At the end of this chapter we characterize sound sources and give a selection of analytical solutions and a variational formulation for the derived governing equations.

In Chapter 3 we describe the developed coupling method in the frequency domain in detail. First, we give a network presentation of the considered acoustical system to motivate the application of model order reduction methods. In the following we explain every step of the coupling method's algorithm in the frame of the CMS, a specific model order reduction method. Thereby we focus on the definition of basis functions, the normal and coupling modes for the pressure and the velocity. These modes are derived out of the velocity potential and approximate the interior sound field of the acoustic system. Further more Hamilton's Principle is applied to compute the solution for the sound field by enforcing minimal energy inside the system. The sound field is presented by employing the Lagrangian of the air, the Lagrangian of the boundary, the virtual work of non-conservative damping forces and the virtual work of the non-conservative forces of sound sources.

In Chapter 4 we analyse numerical methods to compute the normal and the coupling modes. Therein we describe the Spectral Method (SM) and the Spectral Finite Element method (SEM) to gain the normal and the coupling modes for the velocity potential. After a comparison between both methods we propose a strategy to obtain smooth derivatives for the FE solutions of the velocity potential. This strategy is necessary to compute the coupling modes for the pressure and the normal and coupling modes for the velocity.

Chapter 5 includes numerical simulations worked out with the coupling method. These simulation examples demonstrate basic results and exemplify typical applications to room acoustical problems.

To bring this thesis down to a round figure we present an overview of the uncertainty characterization and numerical methods to propagate uncertainty in Chapter 6. Subsequently we specify the sources of uncertainty of the input data related to the coupling method. Then we apply some propagation method approaches to analyse the statistics of the output data.

Finally Chapter 7 summarizes the contributions and conclusions of this research.

2 Acoustical Basics and Governing Equations

Acoustic waves are caused by small oscillations of pressure in a compressible ideal fluid. These oscillations interact in such a way that energy is propagated through the acoustic medium. In the following we start with the governing equations from the fundamental conservation laws for compressible fluids in the time and frequency domain. Next to these basic equations we classify acoustic relevant boundaries in detail. Additionally we define sound sources and acoustic properties. At the end of this chapter we will present some selected analytical solutions and variational solutions for certain assumptions.

2.1 Conservation Laws

Conservation laws are the basic laws to describe phenomena in nature in general. We will use these conservation laws to develop the acoustic Euler equation [64, 94]. After applying a perturbation ansatz we yield the linear Euler equations. These equations are the governing equations for the underlying work.

Theorem 2.1. Reynolds transport theorem
The Reynolds transport theorem is a three-dimensional generalization of the Leibniz integral rule and it is used in formulating conservation laws. For time dependent control volumes $V(t) \subset \mathbb{R}^3$ holds

$$\frac{d}{dt} \int_{V(t)} \Psi(\boldsymbol{x},t) \, d\boldsymbol{x} = \int_{V(t)} \frac{\partial \Psi(\boldsymbol{x},t)}{\partial t} \, d\boldsymbol{x} + \int_{\delta V(t)} \Psi(\boldsymbol{x},t) \frac{\partial \boldsymbol{x}}{\partial t} \, d\boldsymbol{S}, \ \forall \boldsymbol{x} \in V(t),$$

where Ψ is the variable of the system, V is a compact with piecewise smooth boundary, and $d\boldsymbol{S}$ denotes the surface of V.

Theorem 2.2. Conservation of mass
The total mass $m(t)$ of a compact volume $V(t)$ with differentiable boundary is defined by the density ρ as $\int_{V(t)} \rho(x,t) \, d\boldsymbol{x}$ and remains constant in time. The mass cannot be changed as a result of processes acting inside the system, only by external sources of mass

$$\frac{\partial \rho(\boldsymbol{x},t)}{\partial t} + \nabla^T \cdot (\rho(\boldsymbol{x},t)\boldsymbol{v}(\boldsymbol{x},t)) = \frac{\partial m_V(\boldsymbol{x},t)}{\partial t}. \tag{2.1}$$

$\boldsymbol{v}$ denotes the velocity and m_V the mass per volume, injected by sources into the system.

Proof:
With the Reynolds transport Theorem 2.1 and with the Gauss' Theorem follows the integral form of the mass conservation

$$\frac{d}{dt} \int\limits_{V(t)} \rho(\boldsymbol{x},t) \, d\boldsymbol{x} = \int\limits_{V(t)} \frac{\partial \rho(\boldsymbol{x},t)}{\partial t} \, d\boldsymbol{x} + \int\limits_{\delta V(t)} \rho(\boldsymbol{x},t)\boldsymbol{v}(\mathbf{x},t) \cdot \boldsymbol{n} \, d\boldsymbol{S}$$

(2.2)

$$= \int\limits_{V(t)} \frac{\partial \rho(\boldsymbol{x},t)}{\partial t} \, d\boldsymbol{x} + \int\limits_{V(t)} \nabla^T \cdot (\rho(\boldsymbol{x},t)\boldsymbol{v}(\boldsymbol{x},t)) \, d\boldsymbol{x} = 0 \, .$$

$\boldsymbol{v}$ is the differentiable velocity, ρ has to have a local first and second partial derivative in $\boldsymbol{x}$, and $\boldsymbol{n}$ is the outward normal vector at the boundary of $V(t)$. The integral form of the mass conservation (2.2) implies the differential conservation form of the mass conservation also called the continuity equation

$$\frac{\partial \rho(\boldsymbol{x},t)}{\partial t} + \nabla^T \cdot (\rho(\boldsymbol{x},t)\boldsymbol{v}(\boldsymbol{x},t)) = 0 \, .$$

(2.3)

If mass per volume is injected by any external mechanism, the rate at which mass per volume is injected at position $\boldsymbol{x}$ is denoted by $\frac{\partial m_V(\boldsymbol{x},t)}{\partial t}$ and equation (2.3) must be modified to

$$\frac{\partial \rho(\boldsymbol{x},t)}{\partial t} + \nabla^T \cdot (\rho(\boldsymbol{x},t)\boldsymbol{v}(\boldsymbol{x},t)) = \frac{\partial m_V(\boldsymbol{x},t)}{\partial t} \, .$$

Theorem 2.3. Conservation of energy
The total energy $\int_{V(t)} E \, d\boldsymbol{x}$ of a compact volume $V(t)$ with differentiable boundary changes in time at the same rate as the power of the external forces $-\int_{\delta V(t)} p(\boldsymbol{x},t)\boldsymbol{v}(\boldsymbol{x},t) \cdot \boldsymbol{n} \, d\boldsymbol{S}$ resulting in the differential conservation form of the energy conservation

$$\frac{\partial E(\boldsymbol{x},t)}{\partial t} + \nabla^T \cdot (\boldsymbol{v}(\boldsymbol{x},t) \, (E(\boldsymbol{x},t) + p(\boldsymbol{x},t))) = 0 \, .$$

(2.4)

E is the energy and p is the pressure.

Proof:
We have the equivalence of the rate of change of the total energy to the power of external forces

$$\frac{d}{dt} \int\limits_{V(t)} E(\boldsymbol{x},t) \, d\boldsymbol{x} = - \int\limits_{\delta V(t)} p(\boldsymbol{x},t)\boldsymbol{v}(\boldsymbol{x},t) \cdot \boldsymbol{n} \, d\boldsymbol{S} \, .$$

Using Theorem 2.1 yields

$$\int\limits_{V(t)} \frac{\partial E(\boldsymbol{x},t)}{\partial t} \, d\boldsymbol{x} + \int\limits_{\delta V(t)} E(\boldsymbol{x},t)\boldsymbol{v}(\boldsymbol{x},t) \cdot \boldsymbol{n} \, d\boldsymbol{S} = - \int\limits_{\delta V(t)} p(\boldsymbol{x},t)\boldsymbol{v}(\boldsymbol{x},t) \cdot \boldsymbol{n} \, d\boldsymbol{S},$$

(2.5)

and further application of the Gauss' Theorem results in the integral form of the energy conservation

$$\int\limits_{V(t)} \frac{\partial E(\boldsymbol{x},t)}{\partial t}\, d\boldsymbol{x} + \int\limits_{V(t)} \nabla^T \cdot \left(E(\boldsymbol{x},t)\boldsymbol{v}(\boldsymbol{x},t)\right)\, d\boldsymbol{x} \;=\; -\int\limits_{V(t)} \nabla^T \cdot \left(p(\boldsymbol{x},t)\boldsymbol{v}(\boldsymbol{x},t)\right)\, d\boldsymbol{x} \quad (2.6)$$

The integral form of the energy conservation (2.6) implies the differential conservation form of the energy conservation

$$\frac{\partial E(\boldsymbol{x},t)}{\partial t} + \nabla^T \cdot \left(\boldsymbol{v}(\boldsymbol{x},t)\left(E(\boldsymbol{x},t)+p(\boldsymbol{x},t)\right)\right) = 0\,.$$

Theorem 2.4. Conservation of momentum

The momentum $\int_{V(t)} \rho\boldsymbol{v}\, d\boldsymbol{x}$ of a compact volume $V(t)$ with differentiable boundary changes at the same rate as the external forces $-\int_{\delta V(t)} p(\boldsymbol{x},t)\cdot\boldsymbol{n}\, d\boldsymbol{S}$. This yields in the differential conservation form of the momentum

$$\frac{\partial \rho(\boldsymbol{x},t)\boldsymbol{v}(\boldsymbol{x},t)}{\partial t} + \nabla\left(\rho(\boldsymbol{x},t)\boldsymbol{v}(\boldsymbol{x},t)^T\boldsymbol{v}(\boldsymbol{x},t)\right) + \nabla p(\boldsymbol{x},t) = \boldsymbol{f}(\boldsymbol{x},t)\,. \qquad (2.7)$$

f denotes additional external forces.

Proof:
The equivalence of the rate of change of the moment to the external forces writes

$$\frac{d}{dt}\int\limits_{V(t)} \rho(\boldsymbol{x},t)\boldsymbol{v}(\boldsymbol{x},t)d\boldsymbol{x} \;=\; -\int\limits_{\delta V(t)} p(\boldsymbol{x},t)\cdot\boldsymbol{n}\, d\boldsymbol{S}\,.$$

With the Reynolds transport Theorem 2.1 and with the Gauss' Theorem follows the integral form of the momentum conservation

$$\int\limits_{V(t)} \frac{\partial \rho(\boldsymbol{x},t)\boldsymbol{v}(\boldsymbol{x},t)}{\partial t}d\boldsymbol{x} + \int\limits_{\delta V(t)} \rho(\boldsymbol{x},t)\boldsymbol{v}(\boldsymbol{x},t)^T\boldsymbol{v}(\boldsymbol{x},t)\cdot\boldsymbol{n}\, d\boldsymbol{S} \;=\; -\int\limits_{\delta V(t)} p(\boldsymbol{x},t)\cdot\boldsymbol{n}\, d\boldsymbol{S}\,,$$
$$(2.8)$$

$$\int\limits_{V(t)} \frac{\partial \rho(\boldsymbol{x},t)\boldsymbol{v}(\boldsymbol{x},t)}{\partial t}d\boldsymbol{x} + \int\limits_{V(t)} \nabla\left(\rho(\boldsymbol{x},t)\boldsymbol{v}(\boldsymbol{x},t)^T\boldsymbol{v}(\boldsymbol{x},t)\right)\, d\boldsymbol{x} \;=\; -\int\limits_{V(t)} \nabla\left(p(\boldsymbol{x},t)\right)d\boldsymbol{x}\,.$$

The integral form of the momentum conservation (2.6) implies the differential conservation form of the momentum

$$\frac{\partial \rho(\boldsymbol{x},t)\boldsymbol{v}(\boldsymbol{x},t)}{\partial t} + \nabla\left(\rho(\boldsymbol{x},t)\boldsymbol{v}(\boldsymbol{x},t)^T\boldsymbol{v}(\boldsymbol{x},t)\right) + \nabla p(\boldsymbol{x},t) = 0\,. \qquad (2.9)$$

If there are additional external forces $\boldsymbol{f}(\boldsymbol{x}, t)$ acting at all points in V, equation (2.9) turns into

$$\frac{\partial \rho(\boldsymbol{x}, t)\boldsymbol{v}(\boldsymbol{x}, t)}{\partial t} + \nabla \left(\rho(\boldsymbol{x}, t)\boldsymbol{v}(\boldsymbol{x}, t)^T \boldsymbol{v}(\boldsymbol{x}, t) \right) + \nabla p(\boldsymbol{x}, t) = \boldsymbol{f}(\boldsymbol{x}, t).$$

Remark:
Here we assume an inviscid fluid. Otherwise we have to include additional terms on the right side of equation (2.7) to model the viscosity.

Euler gas equations

The three laws of conservation (Theorems 2.2, 2.3, and 2.4) and their differential conservative formulations (2.1), (2.4), and (2.7) form the Euler gas equations in the differential conservative formulation

$$\frac{\partial \boldsymbol{u}(\boldsymbol{x}, t)}{\partial t} + \frac{\partial \boldsymbol{F}_x(\boldsymbol{u}(\boldsymbol{x}, t))}{\partial x} + \frac{\partial \boldsymbol{F}_y(\boldsymbol{u}(\boldsymbol{x}, t))}{\partial y} + \frac{\partial \boldsymbol{F}_z(\boldsymbol{u}(\boldsymbol{x}, t))}{\partial z} = \boldsymbol{s}(\boldsymbol{x}, t), \tag{2.10}$$

with

$$\boldsymbol{u} \;=\; \begin{pmatrix} \rho(\boldsymbol{x},t) \\ \rho(\boldsymbol{x},t)E(\boldsymbol{x},t) \\ \rho(\boldsymbol{x},t)v_x(\boldsymbol{x},t) \\ \rho(\boldsymbol{x},t)v_y(\boldsymbol{x},t) \\ \rho(\boldsymbol{x},t)v_z(\boldsymbol{x},t) \end{pmatrix},$$

$$\boldsymbol{F}_x(\boldsymbol{u}(\boldsymbol{x},t)) \;=\; \begin{pmatrix} \rho(\boldsymbol{x},t)v_x(\boldsymbol{x},t) \\ \rho(\boldsymbol{x},t)v_x(\boldsymbol{x},t)(E(\boldsymbol{x},t)+p(\boldsymbol{x},t)) \\ \rho(\boldsymbol{x},t)v_x^2(\boldsymbol{x},t)+p(\boldsymbol{x},t) \\ \rho(\boldsymbol{x},t)v_x(\boldsymbol{x},t)v_y(\boldsymbol{x},t) \\ \rho(\boldsymbol{x},t)v_x(\boldsymbol{x},t)v_z(\boldsymbol{x},t) \end{pmatrix},$$

$$\boldsymbol{F}_y(\boldsymbol{u}(\boldsymbol{x},t)) \;=\; \begin{pmatrix} \rho(\boldsymbol{x},t)v_y(\boldsymbol{x},t) \\ \rho(\boldsymbol{x},t)v_y(\boldsymbol{x},t)(E(\boldsymbol{x},t)+p(\boldsymbol{x},t)) \\ \rho(\boldsymbol{x},t)v_y(\boldsymbol{x},t)v_x(\boldsymbol{x},t) \\ \rho(\boldsymbol{x},t)v_y^2(\boldsymbol{x},t)+p(\boldsymbol{x},t) \\ \rho(\boldsymbol{x},t)v_y(\boldsymbol{x},t)v_z(\boldsymbol{x},t) \end{pmatrix},$$

$$\boldsymbol{F}_z(\boldsymbol{u}(\boldsymbol{x},t)) \;=\; \begin{pmatrix} \rho(\boldsymbol{x},t)v_z(\boldsymbol{x},t) \\ \rho(\boldsymbol{x},t)v_z(\boldsymbol{x},t)(E(\boldsymbol{x},t)+p(\boldsymbol{x},t)) \\ \rho(\boldsymbol{x},t)v_z(\boldsymbol{x},t)v_x(\boldsymbol{x},t) \\ \rho(\boldsymbol{x},t)v_z(\boldsymbol{x},t)v_y(\boldsymbol{x},t) \\ \rho(\boldsymbol{x},t)v_z^2(\boldsymbol{x},t)+p(\boldsymbol{x},t) \end{pmatrix},$$

$$\boldsymbol{s}(\boldsymbol{x},t) \;=\; \begin{pmatrix} \frac{\partial m_V(\boldsymbol{x},t)}{\partial t} \\ 0 \\ f_x(\boldsymbol{x},t) \\ f_y(\boldsymbol{x},t) \\ f_z(\boldsymbol{x},t) \end{pmatrix}.$$

$\boldsymbol{v} = (v_x, v_y, v_z)^T$ is the velocity and $\boldsymbol{f} = (f_x, f_y, f_z)^T$ the external distributed force. The system has six dependent unknowns and five equations. Consequently we need an additional equation, the state equation, to close the system. It describes the pressure as a function of variables of the fluid. This relation is also known as the compressibility relation between pressure and density for an ideal gas

$$p(\boldsymbol{x},t)\rho(\boldsymbol{x},t)^{-\gamma} = constant, \qquad p(\boldsymbol{x},t) = \frac{1}{\gamma}c^2(\rho,p)\rho(\boldsymbol{x},t). \tag{2.11}$$

which is equivalent to

$$c = \sqrt{\frac{\partial p}{\partial \rho}}. \tag{2.12}$$

Here, γ denotes the heat capacity ratio of the fluid and c the speed of sound.

Linear Euler gas equations

In the following we are interested in the linear phenomena related to pressure, density and velocity. Therefore, only small changes of the primitive variables (see Appendix A.1) are considered, which verify the perturbation ansatz

$$
\begin{aligned}
p(\boldsymbol{x},t) &= p_0 &+& p^*(\boldsymbol{x},t), \\
\boldsymbol{v}(\boldsymbol{x},t) &= \boldsymbol{v}_0 &+& \boldsymbol{v}^*(\boldsymbol{x},t), \\
\rho(\boldsymbol{x},t) &= \rho_0 &+& \rho^*(\boldsymbol{x},t), \\
c(\boldsymbol{x},t) &= c_0 &+& c^*(\boldsymbol{x},t), \\
m_V(\boldsymbol{x},t) &= m_0 &+& m^*(\boldsymbol{x},t), \\
\boldsymbol{f}(\boldsymbol{x},t) &= \boldsymbol{f}_0 &+& \boldsymbol{f}^*(\boldsymbol{x},t),
\end{aligned}
\tag{2.13}
$$

where $p^* \ll p_0$, $\boldsymbol{v}^* \ll \boldsymbol{v}_0$, $\rho^* \ll \rho_0$, $c^* \ll c_0$, $m^* \ll m_0$, and $\boldsymbol{f}^* \ll \boldsymbol{f}_0$. We also assume that the mean flow velocity is zero $\boldsymbol{v}_0 := \boldsymbol{0}$. This means that the other disturbances exist in a stationary medium. Introducing the linearization into the mass equation (2.1) gives

$$
\frac{\partial(\rho_0 + \rho^*(\boldsymbol{x},t))}{\partial t} + \nabla^T \cdot ((\rho_0 + \rho^*(\boldsymbol{x},t))\boldsymbol{v}^*(\boldsymbol{x},t)) = \frac{\partial(m_0 + m^*(\boldsymbol{x},t))}{\partial t}.
$$

This equation holds in absence of perturbations. Neglecting the products of perturbed quantities, we obtain the linearized mass continuity equation for the perturbed quantities

$$
\frac{\partial \rho^*(\boldsymbol{x},t)}{\partial t} + \nabla^T \cdot (\rho_0 \boldsymbol{v}^*(\boldsymbol{x},t)) = \frac{\partial m^*(\boldsymbol{x},t)}{\partial t} = \rho_0 \frac{\partial q^*(\boldsymbol{x},t)}{\partial t}.
\tag{2.14}
$$

Here, q^* denotes the unsteady volume velocity. This procedure is repeated with the conservation of momentum

$$
\frac{\partial((\rho_0 + \rho^*(\boldsymbol{x},t))\boldsymbol{v}^*(\boldsymbol{x},t))}{\partial t} + \nabla\left((\rho_0 + \rho^*(\boldsymbol{x},t))(\boldsymbol{v}^*(\boldsymbol{x},t))^T \boldsymbol{v}^*(\boldsymbol{x},t)\right)
$$

$$
+ \nabla(p_0 + p^*(\boldsymbol{x},t)) = \boldsymbol{f}_0 + \boldsymbol{f}^*(\boldsymbol{x},t).
$$

This equation holds in the absence of perturbations ($\nabla p_0 = \boldsymbol{f}_0$). Neglecting the products of perturbed quantities we end up in the linearized momentum equation

$$
\frac{\partial(\rho_0 \boldsymbol{v}^*(\boldsymbol{x},t))}{\partial t} + \nabla p^*(\boldsymbol{x},t) = \boldsymbol{f}^*(\boldsymbol{x},t).
\tag{2.15}
$$

Finally, we linearize the pressure density relation (2.11). Since the relation (2.11) is valid at any time, it implies first

$$
(p_0 + p^*(\boldsymbol{x},t))(\rho_0 + \rho^*(\boldsymbol{x},t))^{-\gamma} = p_0 \rho_0^{-\gamma}.
\tag{2.16}
$$

After rewriting (2.16) and applying a linearization we yield

$$
p^*(\boldsymbol{x},t) = \left(\gamma \frac{p_0}{\rho_0}\right)\rho^*(\boldsymbol{x},t) = c_0^2 \rho^*(\boldsymbol{x},t).
\tag{2.17}
$$

With (2.14), (2.15), and (2.17) the linear Euler gas equations in pressure and velocity perturbations p^* and $\boldsymbol{v}^*$ write as

$$\frac{\partial \boldsymbol{u}(\boldsymbol{x},t)}{\partial t} + A\frac{\partial \boldsymbol{u}}{\partial x} + B\frac{\partial \boldsymbol{u}}{\partial y} + C\frac{\partial \boldsymbol{u}}{\partial z} = \boldsymbol{s}(\boldsymbol{x},t), \qquad (2.18)$$

with

$$\boldsymbol{u}(\boldsymbol{x},t) = \begin{pmatrix} p^*(\boldsymbol{x},t) \\ v_x^*(\boldsymbol{x},t) \\ v_y^*(\boldsymbol{x},t) \\ v_z^*(\boldsymbol{x},t) \end{pmatrix}, \qquad A = \begin{pmatrix} 0 & \rho_0 c_0^2 & 0 & 0 \\ \frac{1}{\rho_0} & 0 & 0 & 0 \\ 0 & 0 & 0 & 0 \\ 0 & 0 & 0 & 0 \end{pmatrix},$$

$$B = \begin{pmatrix} 0 & 0 & \rho_0 c_0^2 & 0 \\ 0 & 0 & 0 & 0 \\ \frac{1}{\rho_0} & 0 & 0 & 0 \\ 0 & 0 & 0 & 0 \end{pmatrix}, \qquad C = \begin{pmatrix} 0 & 0 & 0 & \rho_0 c_0^2 \\ 0 & 0 & 0 & 0 \\ 0 & 0 & 0 & 0 \\ \frac{1}{\rho_0} & 0 & 0 & 0 \end{pmatrix},$$

$$\boldsymbol{s}(\boldsymbol{x},t) = \begin{pmatrix} c_0^2 \rho_0 \frac{\partial q^*(\boldsymbol{x},t)}{\partial t} \\ \frac{1}{\rho_0} f_x^*(\boldsymbol{x},t) \\ \frac{1}{\rho_0} f_y^*(\boldsymbol{x},t) \\ \frac{1}{\rho_0} f_z^*(\boldsymbol{x},t) \end{pmatrix}.$$

In the following only the pressure and the velocity are formulated directly. The pressure perturbation p^* considers the linear pressure density relation (2.17) and consequently no further formulation is done with the density variable ρ^*.

2.2 State Formulation of the Governing Equations

We will now derive the state formulation for pressure and velocity out of the linear Euler gas equations in Section 2.1 [108].

Once again the linear Euler gas equations are

$$\frac{\partial p^*(\boldsymbol{x},t)}{\partial t} + c_0^2 \nabla^T \cdot (\rho_0 \boldsymbol{v}^*(\boldsymbol{x},t)) = c_0^2 \rho_0 \frac{\partial q^*(\boldsymbol{x},t)}{\partial t}, \tag{2.19}$$

$$\frac{\partial \boldsymbol{v}^*(\boldsymbol{x},t)}{\partial t} + \frac{1}{\rho_0} \nabla p^*(\boldsymbol{x},t) = \frac{1}{\rho_0} \boldsymbol{f}^*(\boldsymbol{x},t). \tag{2.20}$$

Rearranging the linear Euler equations by multiplying (2.20) with ρ_0, building the divergence, and subtracting (2.20), which is multiplied with $\frac{1}{c_0^2}$ and differentiated by t, results in the inhomogeneous wave equation for the pressure variable p^*

$$\Delta p^*(\boldsymbol{x},t) - \frac{1}{c_0^2} \frac{\partial^2 p^*(\boldsymbol{x},t)}{\partial t^2} = \nabla^T \cdot \boldsymbol{f}^*(\boldsymbol{x},t) - \rho_0 \frac{\partial q^*(\boldsymbol{x},t)}{\partial t}. \tag{2.21}$$

Analogously, we divide the Euler equation (2.19) by $\rho_0 c_0^2$, build the gradient and subtract equation (2.20), which is divided by c_0^2 and differentiated by t. This leads to the inhomogeneous wave equation for the velocity variable $\boldsymbol{v}^*$

$$\Delta \boldsymbol{v}^*(\boldsymbol{x},t) - \frac{1}{c_0^2} \frac{\partial^2 \boldsymbol{v}^*(\boldsymbol{x},t)}{\partial t^2} = \nabla \frac{\partial q^*(\boldsymbol{x},t)}{\partial t} - \frac{1}{\rho_0 c_0^2} \frac{\partial \boldsymbol{f}^*(\boldsymbol{x},t)}{\partial t}. \tag{2.22}$$

Boundary conditions

To solve linear hyperbolic partial differential equations of second order such as (2.21) and (2.22), appropriate initial values and boundary conditions are necessary to obtain a well posed problem.

For analysing the needed boundary conditions, we apply the homogeneous relations

$$\frac{\partial p^*(\boldsymbol{x},t)}{\partial t} + c_0^2 \nabla^T \cdot (\rho_0 \boldsymbol{v}^*(\boldsymbol{x},t)) = 0 \quad \text{and} \quad \frac{\partial \boldsymbol{v}^*(\boldsymbol{x},t)}{\partial t} + \frac{1}{\rho_0} \nabla p^*(\boldsymbol{x},t) = \boldsymbol{0}.$$

These relations are out of the homogeneous linear Euler equations (2.19) and (2.20). If we force the variable to follow a certain function at the boundary $\Gamma_D \subset \partial V$, this is called a Dirichlet boundary condition. If we force the normal derivative of the variable at the boundary to satisfy a function, then this is called a Neumann boundary condition on the related boundary $\Gamma_N \subset \partial V$. If we enforce a combination of both boundary conditions on a part of the boundary $\Gamma_R \subset \partial V$, then we have the so-called Robin boundary condition.

Definition 2.1. Boundary conditions in state formulations
In acoustics the following physical boundary conditions are of interest.

$$
\begin{array}{llll}
\text{\textit{open boundary}} & & & \\
\text{\textit{(transmitting)}:} & (\nabla^T \cdot \boldsymbol{v}^*) \cdot \boldsymbol{n} = 0 & \dfrac{\partial p^*}{\partial t} = 0 & (2.23) \\[3mm]
\text{\textit{reflective}} & & & \\
\text{\textit{boundary}:} & \dfrac{\partial \boldsymbol{v}^*}{\partial t}^T \cdot \boldsymbol{n} = 0 & \nabla p^* \cdot \boldsymbol{n} = 0 & (2.24) \\[3mm]
\text{\textit{structural}} & & & \\
\text{\textit{excitation}:} & \dfrac{\partial \boldsymbol{v}^*}{\partial t}^T \cdot \boldsymbol{n} = v_n^* & \nabla p^* \cdot \boldsymbol{n} = -\rho_0 v_n^* & (2.25) \\[3mm]
\text{\textit{impedance}} & & & \\
\text{\textit{boundary}:} & \boldsymbol{v}^{*T} \cdot \boldsymbol{n} = \dfrac{p^*}{Z(\omega)} & \nabla p^* \cdot \boldsymbol{n} = -\dfrac{\rho_0}{Z(\omega)}\dfrac{\partial p^*}{\partial t} & (2.26)
\end{array}
$$

Here, $\boldsymbol{n}$ denotes the outward normal vector at the boundary and $Z(\omega)$ is the impedance function of the impedance boundary.

2.3 Frequency Formulation of the Governing Equations

In this section we derive the governing equations in the frequency domain. The frequency domain is used to describe the domain for analysis of mathematical functions or signals with respect to frequency, rather than time. Later on these governing equations will be the basis for our coupling method in Chapter 3, where we restrict ourselves to time harmonic excitations.

With regard to transient acoustic problems the Fourier transform is applied to transform the function from the time domain into the frequency domain. The Fourier transform and its inverse transform for p^* are defined as

$$p(\boldsymbol{x}, \omega) \;=\; \int\limits_{-\infty}^{+\infty} p^*(\boldsymbol{x}, t) e^{-i\omega t} dt, \qquad p^*(\boldsymbol{x}, t) \;=\; \int\limits_{-\infty}^{+\infty} p(\boldsymbol{x}, \omega) e^{i\omega t} d\omega \,. \qquad (2.27)$$

In (2.27), $p(\boldsymbol{x}, \omega)$ denotes the Fourier transformed of $p^*(\boldsymbol{x}, t)$ and ω stands for the circular frequency. Any time dependent quantity can be written as a sum, respectively as an integral, of time harmonic components. Therefore, there is no loss of generality in linear acoustics by dealing only with time harmonic solutions since any transient solution $p^*(\boldsymbol{x}, t)$ can be reconstructed from a complete spectrum of time harmonic solutions $p(\boldsymbol{x}, \omega)$ and vice-versa. More details about the Fourier transform are given in Appendix A.2.

In the following we restrict ourselves on time harmonic functions $p^*(\boldsymbol{x}, t) = p(\boldsymbol{x}, \omega) e^{i\omega t}$, $\boldsymbol{f}^*(\boldsymbol{x}, t) = \boldsymbol{f}(\boldsymbol{x}, \omega) e^{i\omega t}$ and $q^*(\boldsymbol{x}, t) = q(\boldsymbol{x}, \omega) e^{i\omega t}$. Subsequently the wave equation for the pressure (2.21) becomes the Helmholtz equation in the frequency domain

$$\Delta p(\boldsymbol{x}, \omega) + k^2 p(\boldsymbol{x}, \omega) \;=\; \nabla^T \cdot \boldsymbol{f}(\boldsymbol{x}, \omega) - i\rho_0 \omega q(\boldsymbol{x}, \omega) \,. \qquad (2.28)$$

Here, k denotes the wavenumber, and k_x, k_y and k_z the wavenumbers in x-, y- and z-direction. Between the frequency and the wavenumber the following relation holds

$$\omega^2 = c_0^2 k^2 = c_0^2 \sqrt{k_x^2 + k_y^2 + k_z^2} \,. \qquad (2.29)$$

With $\boldsymbol{v}^*(\boldsymbol{x}, t) = \boldsymbol{v}(\boldsymbol{x}, \omega) e^{i\omega t}$ we apply the same procedure to the wave equation for the velocity variable (2.22) and yield the Helmholtz equation

$$\Delta \boldsymbol{v}(\boldsymbol{x}, \omega) + k^2 \boldsymbol{v}(\boldsymbol{x}, \omega) \;=\; i\omega \nabla q(\boldsymbol{x}, \omega) - \frac{i\omega}{\rho_0 c_0^2} \boldsymbol{f}(\boldsymbol{x}, \omega) \,. \qquad (2.30)$$

Time harmonic assumption

Besides the restriction to linear acoustic problems the second restriction is to focus on time harmonic forces and consequently on time harmonic solutions $p^*(\boldsymbol{x}, t) = p(\boldsymbol{x})e^{i\omega t}$ and $\boldsymbol{v}^*(\boldsymbol{x}, t) = \boldsymbol{v}(\boldsymbol{x})e^{i\omega t}$. The objective is to determine the stationary acoustic field of the pressure $p(\boldsymbol{x})$ and of the velocity $\boldsymbol{v}(\boldsymbol{x})$. The result is complex-valued and the solution of physical interest is the real part of p^* and $\boldsymbol{v}^*$, respectively. For the real part of the pressure we obtain

$$
\begin{aligned}
Re(p^*(\boldsymbol{x}, t)) &= Re((Re(p(\boldsymbol{x})) + i\, Im(p(\boldsymbol{x}))(cos(\omega t) + i\, sin(\omega t))) \\[2mm]
&= |p(\boldsymbol{x})| \left(\frac{Re(p(\boldsymbol{x}))}{|p(\boldsymbol{x})|} cos(\omega t) - \frac{Im(p(\boldsymbol{x}))}{|p(\boldsymbol{x})|} sin(\omega t) \right) \\
& \hspace{9cm} (2.31) \\
&= |p(\boldsymbol{x})| sin(\phi - \omega t),
\end{aligned}
$$

$$
\phi := arctan\left(\frac{Re(p(\boldsymbol{x}))}{Im(p(\boldsymbol{x}))} \right) . \tag{2.32}
$$

Hence, the absolute values of the stationary solutions for pressure and for velocity are the amplitudes of the physical solutions, whereas ϕ is the phase.

Because we are interested in the stationary description of waves reflected on structures, the pressure in the frequency domain can be written as

$$
p(\boldsymbol{x}, \omega) = E_{inc}e^{ik^T\boldsymbol{x}} + E_{ref}e^{-ik^T\boldsymbol{x}} . \tag{2.33}
$$

Here $\boldsymbol{k}$ denotes the wave vector described in detail in Section 2.5. Further on E_{inc} and E_{ref} are complex-valued factors defining the incident and the reflected wave upon the structure. Regarding to [79] we can derive the reflection factor $r = \frac{E_{ref}}{E_{inc}}$, the absorption efficiency $\alpha = 1 - |r|$ used later on for numerical simulation examples in Chapter 6, and the impedance $Z(\omega)$ out of the description (2.33) straight forward.

Remark:
In the following, $p(\boldsymbol{x})$ will denote $p(\boldsymbol{x}, \omega)$ and $\boldsymbol{v}(\boldsymbol{x})$ will denote $\boldsymbol{v}(\boldsymbol{x}, \omega)$ in the frequency domain. This is due to the fact that we restrict ourselves to time harmonic solutions resulting from a harmonic excitation related to the circular frequency ω.

Boundary conditions

To solve the elliptic partial differential equations (2.28) and (2.30), we have to impose appropriate boundary conditions. Thereby, we derive the relation between pressure and velocity (see (2.34)) from the time harmonic case of the homogeneous momentum equation (2.20) as

$$i\omega p(\boldsymbol{x}) + c_0^2 \nabla^T \cdot (\rho_0 \boldsymbol{v}(\boldsymbol{x})) = 0 \quad \text{and} \quad i\omega \boldsymbol{v}(\boldsymbol{x}) + \frac{1}{\rho_0} \nabla p(\boldsymbol{x}) = \boldsymbol{0}. \tag{2.34}$$

Definition 2.2. Boundary conditions in frequency formulations
Common physically boundary conditions occurring in time harmonic acoustics are presented in the following.

open boundary (transmitting):	$(\nabla^T \cdot \boldsymbol{v}) \cdot \boldsymbol{n} = 0$	$p = 0$	(2.35)
reflective boundary:	$\boldsymbol{v} \cdot \boldsymbol{n} = 0$	$\nabla p \cdot \boldsymbol{n} = 0$	(2.36)
structural excitation:	$i\omega \boldsymbol{v} \cdot \boldsymbol{n} = v_n$	$\nabla p \cdot \boldsymbol{n} = -i\rho_0\omega v_n$	(2.37)
structural excitation:	$\boldsymbol{v}^T \cdot \boldsymbol{n} = \dfrac{p}{Z(\omega)}$	$\nabla p \cdot \boldsymbol{n} = -\dfrac{i\rho_0\omega}{Z(\omega)} p$	(2.38)

Besides the reflective (2.36) and the structural excitation boundary condition (2.37), the impedance boundary condition (2.38) is the most realistic one and also the most challenging one. Physical background and detailed information about the impedance will be given in Appendix B.

2.4 Sound Sources and Specific Properties of Sound

Sound sources generate the sound field of interest. We describe the categorization of sound sources and their working mechanism. In addition we introduce some important values to estimate the sound field [32, 33, 79].

Definition 2.3. Characterization of sound sources
There are three different kinds of sound sources related to their mechanisms of sound generation:

  a) *Fluctuating volume/mass displacement/injection,*

  b) *Accelerating/fluctuating force on fluid,*

  c) *Fluctuating fluid shear stress.*

To category **a)** there belong loudspeakers, sirens and exhaust pipe efflux edges amongst others. Their common property is the rate of change of fluid volume displacement, which determines the strength of the sound. Since vibrating surfaces displace fluid volume in an unsteady manner, they fall naturally into category **a)** as well.
Category **b)** sources involve the application of time-varying forces to a fluid without net volume displacement. Examples for this category are the whistling car aerial, turbulence acting on rigid surfaces such as pipe flow-control valves and moving vehicles.
Category **c)** sources produce both zero net volume displacement and zero net force on a fluid. A typical example is the sound made by colliding pool balls. This form of sound generation, which involves no interaction of fluids with solid surfaces, cannot be explained in terms of the linearized equations of inviscid fluid dynamics. It has its origins in the fluctuating shear stresses associated with the turbulent mixing of fluid elements having different time-dependent velocities. For further information we refer to the work of M. J. Lighthill [66].
Most mechanical sources contain a combination of various types of sources. For the mathematical representation category **a)** and **b)** sources will be modelled by elementary sources in the following. The elementary source for **a)** is the pulsating sphere (Figure 2.1) and for **b)** is the oscillating sphere (Figure 2.2). We will give a detailed description for the modelling of sources in Section 3.4.

Figure 2.1: Volume velocity source.

Figure 2.2: Momentum source.

Definition 2.4. Sound energy density
The sound energy density e is the total mechanical energy per unit volume associated with an acoustic disturbance

$$e(x,t) \;=\; T(x,t) \;+\; U(x,t) \;=\; \frac{\rho}{2}|\boldsymbol{v}(x,t)|^2 \;+\; \frac{1}{2\rho c^2}p(x,t)^2,$$

where T is the kinetic energy and U the potential energy.

This is a general definition and holds for all sound fields with small fluctuations and zero mean flow $\boldsymbol{v}_0 = \mathbf{0}$.

Definition 2.5. Sound intensity
The instantaneous sound intensity $\boldsymbol{I}$ describes the local energy flux in sound fields

$$\boldsymbol{I}(\boldsymbol{x},t) \;=\; p(\boldsymbol{x},t)\boldsymbol{v}(\boldsymbol{x},t)\,.$$

The average acoustic intensity in the time interval $[0,T]$ for a plane wave is

$$\boldsymbol{I}_n(\boldsymbol{x}) \;=\; \frac{1}{T}\int\limits_0^T p(\boldsymbol{x},t)\boldsymbol{v}_n(\boldsymbol{x},t)\,dt,$$

where the subscript n indicates the direction of the intensity, the average direction of the energy.

Hence, if there is no external source the relation $\nabla^T\mathbf{I} = -\frac{\partial e}{\partial t}$ holds.

Definition 2.6. Mean sound power
The mean sound power P is the sound intensity of a sound source. The sound power describes the source strength

$$P \;=\; \int_S I_n(\boldsymbol{x})\,dS\,.$$

I_n *denotes the mean axial intensity in outward normal direction to the surface of interest S for stationary sound fields.*

For example, the sound power of a speaking person is around $10^{-5}W$. It is an important value to characterize a sound source or a sound field, because it is independent of the position of the sound source.

2.5 Selected Solutions for the Wave Equation and the Helmholtz Equation

In the following, we will present some results based on analytical and variational solutions. Thereby, we consider solutions for the wave equation in the time domain and for the Helmholtz equation in the frequency domain [90, 104].

2.5.1 Analytical Solutions

Wave equation – harmonic plane waves in free space

The most simple case are plane waves in an unbounded three dimensional domain, where no boundary conditions are necessary. They have a constant frequency and their wave fronts are parallel planes of constant amplitudes normal to the phase velocity vector

$$p^*(x,t) = Re\left(Ae^{i(\omega t - k^T x)} + Be^{i(\omega t + k^T x)}\right), \qquad A, B \in \mathbb{C}, \; x \in \mathbb{R}^3, \qquad (2.39)$$

with

$$k = k_x e_1 + k_y e_2 + k_z e_3, \qquad k = |k| = \sqrt{k_x^2 + k_y^2 + k_z^2} = \frac{\omega}{c_0}.$$

The wave vector k describes the direction of wave propagation, where $\{e\}_j$ are the unit vectors in the Cartesian coordinate system, and k is the related wavenumber. With the homogeneous Euler gas equation (2.20) and with (2.39) follows the related solution for the sound velocity

$$v^*(x,t) = Re\left(\frac{k}{\rho_0 i \omega}\left(Ae^{i(\omega t - k^T x)} - Be^{i(\omega t + k^T x)}\right)\right). \qquad (2.40)$$

The solution is not unique. Hence, additional constraints are necessary.

Wave equation – initial condition

The general solution of the wave equation (2.21) in the homogeneous case for one spatial dimension $x \in \mathbb{R}$ is a sum of two functions

$$p^*(x,t) = g(x - c_0 t) + f(x + c_0 t), \quad f, g \in C^2.$$

19

g is a wave going from the left to the right and f is going from the right to the left. With initial conditions

$$p^*(x,0) = \phi(x),$$
$$\frac{\partial p^*(x,t)}{\partial t}\Big|_{t=0} = \psi(x),$$

follows the solution known as D'Alembert's formula

$$p^*(x,t) = \frac{1}{2}\left(\phi(x - c_0 t) + \phi(x + c_0 t) + \frac{1}{c_0}\int_{x-c_0 t}^{x+c_0 t} \psi(\xi)d\xi\right) \tag{2.41}$$

Remark:
We achieve the solution in two and three dimensions with the Poisson formula and the Kirchhoff formula [90, 104].

Helmholtz equation – reflective rectangular volume

Problems for general enclosures have no analytical solution for all kinds of boundaries. Only simple geometries like cuboids, spheres, and cylinders have analytical solutions for special boundary conditions and no present sources. We assume the closed rectangular cavity $V = [0, L_x] \times [0, L_y] \times [0, L_z]$. In this case the inhomogeneous Helmholtz equation for the pressure (see (2.28)) reduces to $\Delta p(\boldsymbol{x}) + k^2 p(\boldsymbol{x}) = 0$ for $\boldsymbol{x} \in V$, which is an eigenvalue problem for the Helmholtz operator $\Delta + k^2$. We choose the method separation of variables, which is possible because of the geometry and linear homogeneous boundary conditions.

$$p(\boldsymbol{x}) = p_x(x)p_y(y)p_z(z).$$

For the homogeneous Helmholtz equation this yields

$$\frac{\partial^2 p_x}{\partial x^2}p_y p_z + p_x\frac{\partial^2 p_y}{\partial y^2}p_z + p_x p_y\frac{\partial^2 p_z}{\partial z^2} + (k_x^2 + k_y^2 + k_z^2)p_x p_y p_z = 0,$$

and finally

$$\frac{\frac{\partial^2 p_x}{\partial x^2}}{p_x} + \frac{\frac{\partial^2 p_y}{\partial y^2}}{p_y} + \frac{\frac{\partial^2 p_z}{\partial z^2}}{p_z} + (k_x^2 + k_y^2 + k_z^2) = 0.$$

This equation can be separated into three ordinary differential equations. The reflective boundary condition (2.36) induces in the conditions $\frac{\partial p_x(x)}{\partial x} = 0$ for $x \in \{0, L_x\}$, $\frac{\partial p_y(y)}{\partial y} = 0$ for $y \in \{0, L_y\}$, and $\frac{\partial p_z(z)}{\partial z} = 0$ for $z \in \{0, L_z\}$. The solution for the eigenvalues, resulted from the application of the method separation of variables, is an unbounded discrete spectrum and the solution for the pressure is

$$p(\boldsymbol{x}) = \sum_l \sum_m \sum_n \epsilon_{lmn} p_{lmn}(\boldsymbol{x}), \qquad \epsilon_{lmn} \in \mathbb{C}, \ l, m, n \in \mathbb{Z},$$

$$= \sum_l \sum_m \sum_n \epsilon_{lmn} cos(k_l x) cos(k_m y) cos(k_n z), \qquad (2.42)$$

with

$$k_l = \frac{l\pi}{L_x}, \quad k_m = \frac{m\pi}{L_y}, \quad k_n = \frac{n\pi}{L_z},$$

$$\mathbf{k}_{lmn} = k_l e_1 + k_m e_2 + k_n e_3, \qquad k_{lmn} = \sqrt{k_l^2 + k_m^2 + k_n^2}. \qquad (2.43)$$

$\omega_{lmn} = c_0 k_{lmn}$ are the circular eigenfrequencies and $p_{lmn}(\boldsymbol{x}) = cos(k_l x) cos(k_m y) cos(k_n z)$ are the related eigenmodes of the cavity. Distinct eigenmodes are orthogonal inside of the cavity. Using equation (2.34) we get the solution for the velocity

$$\boldsymbol{v}(\boldsymbol{x}) = \sum_l \sum_m \sum_n -\frac{\epsilon_{lmn}}{\rho_0 \omega_{lmn}} \begin{pmatrix} k_l \, sin(k_l x) \, cos(k_m y) \, cos(k_n z) \\ k_m \, cos(k_l x) \, sin(k_m y) \, cos(k_n z) \\ k_n \, cos(k_l x) \, cos(k_m y) \, sin(k_n z) \end{pmatrix}. \qquad (2.44)$$

We are aware of, that the velocity fulfills the hard surface boundary condition (2.36) for every component as expected.

Helmholtz equation – rectangular volume with impedance boundary

There are no analytical solutions for general impedance boundary conditions, even for simple geometries. But forcing some assumptions enables us to find some approximated solutions for special cases.

We consider again the geometry $V = [0, L_x] \times [0, L_y] \times [0, L_z]$ assuming a reflective boundary condition (2.36) everywhere except at $z = L_z$. There we enforce the homogeneous Robin boundary condition

$$\alpha p(\boldsymbol{x}) + \beta \frac{\partial p(\boldsymbol{x})}{\partial \boldsymbol{n}} = 0, \qquad \forall \boldsymbol{x} \in \delta V, \ z = L_z, \ \alpha, \beta \in \mathbb{C}, \qquad (2.45)$$

which is comparable to the impedance boundary in (2.38). The separation of variables $p(\boldsymbol{x}) = p_x(x) p_y(y) p_z(z)$ allows to treat each direction separately. For the x- and y-direction

21

there yield the same solution such as in the previous example. With the reflective boundary at $z = 0$ follows

$$p_z(z) = \tilde{A}_z cos(k_n z).$$

With $\tilde{Z} := \frac{\alpha}{\beta}$ and the impedance boundary condition we get

$$\tilde{Z}\frac{\partial p_z(z)}{\partial z} + p_z(z) = 0,$$

and consequently

$$\tilde{Z}k_n tan(k_n L_z) = 1.$$
(2.46)

Considering a high value for $\tilde{Z}$, that means a nearly reflecting boundary, results in

$$k_n L_z = n\pi + \epsilon, \quad n \in \mathbb{Z}, 0 < \epsilon \ll 1.$$

Finally, with $tan(k_n L_z) \approx \epsilon$ yields

$$k_n = \frac{1}{2L_z}\left(n\pi + \sqrt{(n\pi)^2 + \frac{4L_z}{\tilde{Z}}}\right).$$
(2.47)

If we assume a nearly totally absorbing impedance, that means $\tilde{Z} \ll 1$, the solution for k_n is

$$k_n = \frac{1}{L_z + \tilde{Z}}\left(\frac{\pi}{2} + n\pi\right).$$
(2.48)

The pressure p with the related expression for k_n (see (2.47) and (2.48)) results in

$$
\begin{aligned}
p(\boldsymbol{x}) &= p_x(x)p_y(y)p_z(z) \\[2mm]
&= \sum_l \sum_m \sum_n \epsilon_{lmn} cos(k_l x)cos(k_m y)cos(k_n z), \quad \epsilon_{lmn} \in \mathbb{C}, \, l,m,n \in \mathbb{Z},
\end{aligned}
$$
(2.49)

with

$$k_l = \frac{l\pi}{L_x}, \quad k_m = \frac{m\pi}{L_y}, \quad l,m \in \mathbb{Z}.$$

Remark:
The method of separating the variables allows us to solve the Helmholtz problem also for other symmetric geometries like spheres and cylinders with simple boundary conditions.

Modal density

For higher frequencies of excitation the density of eigenfrequencies in rectangular and in general closed cavities is increase, which is shown in Figure 2.3 for a rectangular volume. The increasing modal density is typical for all volumes. This implies a great challenge in Technical Acoustics. For higher frequencies the absolute difference between eigenfrequencies decreases and consequently it is more difficult to detect and to separate single eigenfrequencies. In this case the response of the sound field becomes very sensitive to the point of excitation, amongst others. Therefore it is necessary to refine the spatial resolution for higher frequencies, which results in increasing computational costs. That is the reason why local discretization methods like the FE method are wide spread in the low-frequency range $(0-800\,Hz)$ but not in the mid- or high-frequency range $(\geq 800\,Hz)$.

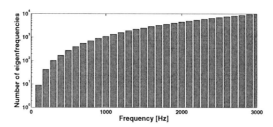

Figure 2.3: Increasing modal density for $L_x = 6\,m$, $L_y = 3\,m$ and $L_z = 2\,m$.

2.5.2 Variational Solutions

Besides the analytical solutions, also called strong or classical solutions, from Section 2.5.1 we are interested in the weak or variational solution of the Helmholtz equation.

We take the Helmholtz equation (2.28) with a zero right-hand side on the bounded domain V with the boundary $\delta V = \overline{\Gamma_D \cup \Gamma_N \cup \Gamma_R}$ and $\Gamma_D \cap \Gamma_N = \Gamma_N \cap \Gamma_R = \Gamma_R \cap \Gamma_D = \emptyset$:

$$\Delta p(\boldsymbol{x}) + k^2 p(\boldsymbol{x}) = 0, \qquad \forall\, \boldsymbol{x} \in V,\ p(\boldsymbol{x}) \in C^2(V) \cap C(\overline{V}),$$

$$p(\boldsymbol{x}) = p_D(\boldsymbol{x}), \qquad \forall\, \boldsymbol{x} \in \Gamma_D,$$
$$\frac{\partial p(\boldsymbol{x})}{\partial n} = -i\rho_0 c_0 k v_n(\boldsymbol{x}), \qquad \forall\, \boldsymbol{x} \in \Gamma_N,$$
$$\frac{\partial p(\boldsymbol{x})}{\partial n} = -i\rho_0 c_0 k v_n(\boldsymbol{x}) p(\boldsymbol{x}), \quad \forall\, \boldsymbol{x} \in \Gamma_R.$$

All relevant acoustic conditions from Section 2.3 are covered (see Definition 2.2) in this example. After multiplication with the test functions $\nu \in C_0^\infty(V)$ (here the subscript 0

23

denotes that ν vanishes at the boundary of V) we integrate over the domain V. Using Green's formula yields

$$\int\limits_V -\nabla p(\boldsymbol{x})\nabla\overline{\nu}(\boldsymbol{x}) + k^2\,p(\boldsymbol{x})\overline{\nu}(\boldsymbol{x})\,d\boldsymbol{x} - \int\limits_{\Gamma_R} i\rho_0 c_0 k v_n(\boldsymbol{x})p(\boldsymbol{x})\overline{\nu}(\boldsymbol{x})\,ds = \int\limits_{\Gamma_N} i\rho_0 c_0 k v_n(\boldsymbol{x})\overline{\nu}(\boldsymbol{x})\,ds\,.$$

The suitable function spaces are

$$p(\boldsymbol{x}) \in H_D^1(V) \;\; := \;\; \{p(\boldsymbol{x}) \in H^1(V),\; p(\boldsymbol{x}) = p_D\;\forall\boldsymbol{x}\in\Gamma_D\}\,,$$
$$\nu(\boldsymbol{x}) \in H_0^1(V) \;\; := \;\; \{\nu(\boldsymbol{x}) \in H^1(V),\; \nu(\boldsymbol{x}) = 0\;\;\forall\boldsymbol{x}\in\Gamma_D\}\,.$$

In the language of linear forms we can state the weak form of the Helmholtz problem in the bounded domain V:
find a function $p \in H_D^1(V)$ such that

$$a(p,\nu) \;\; = \;\; l(\nu), \qquad\qquad \forall\nu(\boldsymbol{x}) \in H_0^1(V), \qquad\qquad (2.50)$$
$$(2.51)$$

with the sesquilinear form $a(\cdot,\cdot): H_D^1(V) \times H_0^1(V) \to \mathbb{C}$ being defined by

$$a(p,\nu) \;\; = \;\; \int\limits_V -\nabla p(\boldsymbol{x})\nabla\overline{\nu}(\boldsymbol{x}) + k^2 p(\boldsymbol{x})\overline{\nu}(\boldsymbol{x})\,d\boldsymbol{x} - \int\limits_{\Gamma_R} i\rho_0 c_0 k v_n(\boldsymbol{x})p(\boldsymbol{x})\overline{\nu}(\boldsymbol{x})\,ds,$$

and the continuous linear form $l \in H_D^1(V)$

$$l(\nu) \;\; = \;\; \int\limits_{\Gamma_N} i\rho_0 c_0 k v_n(\boldsymbol{x})\overline{\nu}(\boldsymbol{x})\,ds\,.$$

The space H_D^1 is equipped with the L^2 norm and the H^1 semi-norm

$$\|p\|_0^2 = \int\limits_V p(\boldsymbol{x})\overline{p}(\boldsymbol{x})\,d\boldsymbol{x} \quad\text{and}\quad |p|_1^2 = \int\limits_V \nabla p(\boldsymbol{x})\nabla\overline{p}(\boldsymbol{x})\,d\boldsymbol{x}\,.$$

With the idea of I. Babuska [3] and F. Brezzi [10] the *inf-sup* stability condition for the variational conditions and approximate numerical solutions is necessary. This condition reads as

$$\vartheta = \inf_{p\in H_D^1(V)} \sup_{\nu\in H_0^1(V)} \frac{|a(p,\overline{\nu})|}{|p|_{H_D^1(V)}|\overline{\nu}|_{H_0^1(V)}}\,.$$

Theorem 2.5. Existence and uniqueness
Let $a(\cdot,\cdot): H_D^1(V) \times H_0^1(V) \to \mathbb{C}$ be the sesquilinear form defined by the equation (2.50). Then the inf-sup condition is of order k^{-1} and there exist positive constants C_1 and C_2, independent on k, such that

$$\frac{C_1}{k} \le \vartheta \le \frac{C_2}{k}\,.$$

Proof: See [83].

3 Modal-Based Fluid Structure Coupling Method

In the following we present the coupling method in the frequency domain. Therein we focus on the basis functions and the models for realistic boundary conditions. Finally, we will discuss the application of the method in the presence of simple sound sources.

3.1 Coupling Method in the Frequency Domain

In this first section we subdivide the closed acoustic system of interest into its components. These components interact with each other and represent an acoustic network, on which we will apply the coupling method in Section 3.1.2.

3.1.1 Acoustic Network Approach

It is common practice to subdivide large-scale and complex structures into several smaller components or substructures for the purpose of analysis [72, 93, 96]. This allows to analyse a complex system by looking at simpler subsystems and components with less degrees of freedom. In Figure 3.1 we present a 2d acoustic network that can be used to model a room or a car interior, for example.

The topology of the acoustic network shows the structure of the air components, related sound sources, and the boundary components. The latter build the enclosure of the acoustic cavity and divide into two types the reflective boundaries and the absorbers. Reflective boundaries are acoustically hard surfaces, which do not allow transmission or absorption of sound waves. The more realistic impedance boundaries are modelled by absorbers. They reflect, transmit, and absorb sound waves at the same time. We give a detailed description about impedance boundaries in the Appendix B. Its mechanism is considered by the absorber's specific frequency dependent impedance function $Z(\omega)$. We will describe the pre-process of the impedance function in Section 3.3.

We remark, that between two air components there is an acoustically soft boundary, where sound waves are completely transmitted from one component to the other.

Besides the air components and boundary components there exist sound source components in the acoustic network representation. In Section 2.4, we have given a short characterization of sound sources and their mechanisms of sound generation. Each sound source has its distinct air component, where it is located. We will discuss the modelling of these sound sources in Section 3.4.

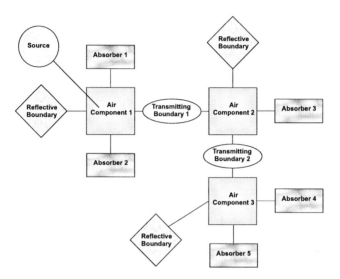

Figure 3.1: Topology of an acoustic network of a closed cavity subdivided into air components including sound sources and different types of boundaries.

Amongst other room acoustic phenomena of interest, we focus on the interior sound field of the acoustic cavity. This sound field is defined by the pressure or velocity distribution. These distributions in turn are influenced by the geometry and the boundaries of the cavity as well as the type and position of sound sources. How we account for all these influences of the interior sound field will be the topic of Section 3.1.2.

Table 3.1: Summary of the acoustic network components and there functionality

Symbol	Description
	air component a part of the total air volume V
	totally reflecting boundary component
	totally transmitting boundary component located between two air components
	transmitting, absorbing and reflecting boundary component
	harmonically oscillating sound source component located in one air component

3.1.2 Fluid-structure interaction Coupling Method

The coupling method is based on the Component Mode Synthesis (CMS) [77]. The CMS is a well established method for efficiently constructing models to analyse and predict the overall structural dynamic response. If substructures are coupled, then appropriate dynamic boundary conditions have to be satisfied at the interfaces. Typically, the solution within each substructure is approximated by a set of basis vectors (Ritz vectors), so-called modes, where the number of vectors is substantially smaller than the number of physical degrees of freedom (dofs). The approximation of the solution within the substructure using numerical methods provides a global approximation of the global system.

We describe the sound field inside a closed cavity V by the pressure p^* and the velocity $\boldsymbol{v}^*$. In the following, the coupling method is applied to one single air component (see Figure 3.1) for simplicity. The basic idea of the CMS is to divide the modes, that constitute the basis functions for the state variables pressure and velocity, into two sets.

Definition 3.1. Superposition of normal and coupling modes
The solution for the pressure p^* and for the velocity $\boldsymbol{v}^*$ is restricted to time harmonic excitation and is composed of two sums of modes

$$
\begin{aligned}
p^*(\boldsymbol{x},t) &= \sum_r p_r^N(\boldsymbol{x})\left(A_r e^{i\omega t} + \overline{A_r}e^{-i\omega t}\right) + \sum_s p_s^C(\boldsymbol{x})\left(B_s e^{i\omega t} + \overline{B_s}e^{-i\omega t}\right), \quad A_r, B_s \in \mathbb{C}, \\
&\qquad\qquad\qquad\qquad\qquad\qquad\qquad\qquad\qquad\qquad\qquad\qquad\qquad\qquad\qquad\qquad (3.1) \\
\boldsymbol{v}^*(\boldsymbol{x},t) &= \sum_r \boldsymbol{v}_r^N(\boldsymbol{x})\left(A_r e^{i\omega t} - \overline{A_r}e^{-i\omega t}\right) + \sum_s \boldsymbol{v}_s^C(\boldsymbol{x})\left(B_s e^{i\omega t} - \overline{B_s}e^{-i\omega t}\right).
\end{aligned}
$$

Here, ω denotes the circular frequency of excitation, related to the sound source. The superscript N stands for the normal modes and the superscript C for the coupling modes. $\overline{A_r}$ and $\overline{B_s}$ represent the complex conjugated values of A_r and B_s, respectively.

Remark:

The normal modes are the eigenfunctions for the air component with totally reflecting boundary conditions. They are also called rigid-wall eigenfunctions. The coupling modes p^C and $\boldsymbol{v}^C$ enable the coupling to a specific boundary or an adjacent air component inside the cavity. In Chapter 3.2 we will discuss the types of normal and coupling modes and in Chapter 4 we will introduce the numerical methods to compute the modes. We need the approach of $e^{i\omega t}$ and $e^{-i\omega t}$ to achieve a phase shift in ω the frequency of excitation. All modes are computed by means of the velocity potential Φ.

Definition 3.2. The velocity potential
The velocity potential Φ^* with the relation $\boldsymbol{v}^* = \nabla\Phi^*$ fulfills the wave equation (2.21). Φ^* underlies the time harmonic excitation and ϕ fulfills the Helmholtz equation (2.28)

$$
\begin{aligned}
\Delta\Phi^*(\boldsymbol{x},t) &= \tfrac{1}{c_0^2}\tfrac{\partial^2\Phi^*(\boldsymbol{x},t)}{\partial t^2}, & \Phi^*(\boldsymbol{x},t) &= \Phi(\boldsymbol{x})e^{i\omega t}, \\
&&&\qquad\qquad (3.2) \\
\Delta\Phi(\boldsymbol{x}) &= -k^2\Phi(\boldsymbol{x}), & \omega^2 &= c_0^2 k^2.
\end{aligned}
$$

For the normal modes there arises an eigenvalue problem. Due to the assumption of totally reflective boundaries we yield in

$$
\begin{aligned}
\Delta\Phi_r^N(\boldsymbol{x}) + k_r^2\Phi_r^N(\boldsymbol{x}) &= 0, & r &= 1,2,\dots & (3.3) \\
\nabla\Phi_r^N(\boldsymbol{x})\cdot\boldsymbol{n} &= 0, & \forall \boldsymbol{x} &\in \Gamma_N = \partial V.
\end{aligned}
$$

Γ_N denotes the Neumann boundary of the volume (see Section 2.2). For the coupling modes we have to consider an inhomogeneous boundary condition in the Helmholtz equation. Consequently, we get a linear system of equations for each Φ_s^C.

We define the modes for the pressure and the velocity out of the normal and coupling modes for the velocity potential.

Definition 3.3. Basis functions
The basis functions for the coupling method, the pressure normal modes p_r^N and coupling modes p_s^C as well as the velocity normal modes v_r^N and coupling modes v_s^C, are defined by the velocity potential modes as follows

$$
\begin{aligned}
p_r^N(\boldsymbol{x}) &:= -\tfrac{\rho_0 c_0^2}{\omega} \Delta \Phi_r^N(\boldsymbol{x}), & p_s^C(\boldsymbol{x}) &:= -\tfrac{\rho_0 c_0^2}{\omega} \Delta \Phi_s^C(\boldsymbol{x}), \\
v_r^N(\boldsymbol{x}) &:= i \nabla \Phi_r^N(\boldsymbol{x}), & v_s^C(\boldsymbol{x}) &:= i \nabla \Phi_s^C(\boldsymbol{x}).
\end{aligned}
\tag{3.4}
$$

Φ_r^N denote the velocity potential normal modes and Φ_s^C the velocity potential coupling modes.

Theorem 3.1.
The solutions for pressure and velocity have to satisfy the homogeneous linear Euler equation (2.19) and the linear Euler equation (2.20) containing the source term $\boldsymbol{f}^*$

$$
\frac{\partial p^*(\boldsymbol{x},t)}{\partial t} = -\rho_0 c_0^2 \nabla^T \boldsymbol{v}^*(\boldsymbol{x},t), \tag{3.5}
$$

$$
\rho_0 \frac{\partial \boldsymbol{v}^*(\boldsymbol{x},t)}{\partial t} + \nabla p^*(\boldsymbol{x},t) = \boldsymbol{f}^*(\boldsymbol{x}). \tag{3.6}
$$

Remark:

In the following lines we show, that with the definition of the modes (3.4) and with the chosen basis functions (3.1), equation (3.5), also called Newton's law, is fulfilled

$$
\begin{aligned}
\frac{\partial p^*(\boldsymbol{x},t)}{\partial t} &= \\
\frac{\partial}{\partial t} &\left(\sum_r p_r^N(\boldsymbol{x}) \left(A_r e^{i\omega t} + \overline{A_r} e^{-i\omega t} \right) + \sum_s p_s^C(\boldsymbol{x}) \left(B_s e^{i\omega t} + \overline{B_s} e^{-i\omega t} \right) \right) \\
&= -i\rho_0 c_0^2 \left(\sum_r \Delta\Phi_r^N(\boldsymbol{x}) \left(A_r e^{i\omega t} - \overline{A_r} e^{-i\omega t} \right) + \sum_s \Delta\Phi_s^C(\boldsymbol{x}) \left(B_s e^{i\omega t} - \overline{B_s} e^{-i\omega t} \right) \right) \\
&= -\rho_0 c_0^2 \nabla^T \left(\sum_r v_r^N(\boldsymbol{x}) \left(A_r e^{i\omega t} - \overline{A_r} e^{-i\omega t} \right) + \sum_s v_s^C(\boldsymbol{x}) \left(B_s e^{i\omega t} - \overline{B_s} e^{-i\omega t} \right) \right) \\
&= -\rho_0 c_0^2 \nabla^T \boldsymbol{v}^*(\boldsymbol{x},t).
\end{aligned}
$$

The coefficients A_r and B_s belonging to the basis functions are determined with Hamilton's Principle, the core of the coupling method, so that the solution satisfies (3.6).

Hamilton's Principle

Hamilton's Principle states that the dynamics of a system are determined by a variational problem for a functional based on the Lagrangian. This Lagrangian contains all physical information of the system and the forces acting on it. In the following we will apply Hamilton's Principle. Therefore we vary the coefficients A_r and B_s of the solution and compute the final solution by forcing a minimum of energy in the system. One part of Hamilton's Principle is the Lagrangian for the air volume including the kinetic energy and the potential energy, which we have already introduced in Section 2.4.

Definition 3.4. Lagrangian of the energy in the air
The Lagrangian of the air L_A is defined by

$$L_A(t) := T_A(t) - U_A(t) \tag{3.7}$$

with the kinetic energy

$$T_A(t) := \frac{\rho_0}{2} \int_V |v^*(x, t)|^2 \, dV \tag{3.8}$$

and the potential energy

$$U_A(t) := \frac{1}{2\rho_0 c_0^2} \int_V |p^*(x, t)|^2 \, dV . \tag{3.9}$$

The subscript A denotes the air inside the acoustic volume.

Furthermore harmonically oscillating loads of sound sources are considered in Hamilton's Principle by their virtual work.

Definition 3.5. Virtual work of the sound source
The virtual work of non-conservative forces induced by the sound source is defined by

$$\delta W_{Load}^{nc}(t) := \int_{A_{Load}} p_{Load} \, n_{A_{Load}}(x) \, \delta w^*(x, t) \, dA . \tag{3.10}$$

The index nc stands for non-conservative, $\delta w^*(x, t)$ denotes the virtual displacement underlying the time harmonic assumption and A_{Load} is the surface of the sound source specified by the normal vector $n_{A_{Load}}(x)$.

We give a detailed discussion of sound sources in Section 3.4.
Finally, we consider the influence of realistic boundaries by the Lagrangian function L_{BC} and by the virtual work of the non-conservative damping forces δW_{BC}^{nc}. Both are dependent on the specific boundary related impedance Z. We will describe the computation of these terms with the help of the coupling modes in Section 3.3. After including both boundary related terms L_{BC} and δW_{BC}^{nc} the final formulation of energy equilibrium using

Hamilton's Principle yields

$$\min_{T>0} \int_0^T L_A(t) + L_{BC}(t,Z) + \delta W_{BC}^{nc}(t,Z) + \delta W_{Load}^{nc}(t) \, dt \, . \tag{3.11}$$

Remark:

Hamilton's Principle requires that the first-order change of the integral in (3.11) is zero for all possible perturbations. In our case we have no functional and consequently we vary not the functions but the coefficients A_r and B_s. The terms L_A, L_{BC}, δW_{BC}^{nc}, and δW_{Load}^{nc} are dependent on quadratic or mixed products of the unknown coefficients A_r and B_s only. Consequently we can built up a linear system of equations for the coefficients. We describe the construction of the linear system of equations in detail in Section 3.1.3. Because of the time harmonic assumption we choose $T = \frac{2\pi}{\omega}$ in equation (3.11).

3.1.3 Implementation of the Coupling Method

To realize the minimum of energy in equation (3.11) with Hamilton's Principle numerically, we generates a linear matrix vector equation for the coefficients A_r and B_s. With a finite number of normal and coupling modes we can approximate the pressure and the velocity in the air volume as in (3.1).

Theorem 3.2. Solution for the discrete minimum of energy
Approximating the solution of (3.11) with a finite number of modes for the pressure p and for the velocity v yields the following linear system of equations with the unknown coefficients $A_1, ..., A_{N_r}$ for the normal modes and $B_1, ..., B_{N_s}$ for the coupling modes

$$\boldsymbol{Mc} \;=\; \boldsymbol{f}_{Load}. \tag{3.12}$$

We use the notation

$$M_{ij} \;=\; \frac{\partial}{\partial c_j}\left(\frac{\partial}{\partial \overline{c_i}}\left(L_A + L_{BC}(Z)\right) + \frac{\partial}{\partial \overline{\delta c_i}}\delta W_{BC}^{nc}(Z)\right),$$

$$\delta \boldsymbol{c} \;=\; [\delta A_1, \delta A_2, ..., \delta A_{N_r}, \delta B_1, \delta B_2, ..., \delta B_{N_s}]^T,$$

$$\boldsymbol{c} \;=\; [A_1, A_2, ..., A_{N_r}, B_1, B_2, ..., B_{N_s}]^T,$$

$$\boldsymbol{f}_{Load\, i} \;=\; -\frac{\partial}{\partial \overline{\delta c_i}}\delta W_{Load}^{nc}.$$

δA_r and δB_s denote the virtual coefficients belonging to the virtual quantities. $\overline{\delta c}$ and $\overline{c}$ notice the complex conjugated quantities of δc and c, respectively.

Proof:
The kinetic and the potential energies, (3.8) and (3.9), which build the Lagrangian of the air L_A, consist of quadratic terms of the pressure and the velocity. Because the pressure and the velocity consist of a finite number of normal and coupling modes, the Lagrangian is composed of quadratic and mixed terms of the coefficients A_r, $\overline{A_r}$, B_s, and $\overline{B_s}$. After the integration there remain mixed terms only, because of the time harmonic assumption ($T = \frac{2\pi}{\omega}$). Consequently the first-order variation of the Lagrangian in the complex conjugated coefficients yields in linear terms of the coefficients. The same holds for the Lagrangian of the boundary L_{BC}. The virtual work of the non-conservative damping forces δW_{BC}^{nc} consist of mixed terms of coefficients and virtual coefficients δA_r and δB_s. After varying in the virtual coefficients we end up in linear terms of the coefficients again. We analyse the construction of L_{BC} and δW_{BC}^{nc} in detail in Section 3.3. The virtual work of non-conservative forces is composed of linear terms in the virtual coefficients and builds the right side of the linear system of equations (3.12) after varying in the δA_r and δB_s.

Remark:

In equation (3.12) M and L_A are regular symmetric matrices. L_{BC} and W_{BC}^{nc} are matrices with zero components except one diagonal block related to the coupling modes, which are responsible for the coupling to the related absorber component (see Section 3.3). The resulting solution coefficients give the frequency response for the specific geometry, boundary and load configuration, a fixed number of modes, and a chosen circular frequency of excitation ω. For an overview of all steps of the algorithm we present a flowchart in Figure 3.2. For the reason of efficiency we compute the normal and coupling modes as a pre-processing step for a fixed geometry and adapt them with the related material parameter and the frequency of excitation (see equation (3.4)).

If there is no sound source ($f_{Load} = 0$), we rearrange the homogeneous system $Mc = 0$ only for special cases of impedance boundaries into a general eigenvalue problem in ω^2. For these cases the eigenvectors of the acoustic volume are a combination of the basis functions, the normal modes and the coupling modes. For general configurations of boundaries the boundary related impedance Z is nonlinear. Consequently we cannot convert the system explicitly into an eigenvalue problem.

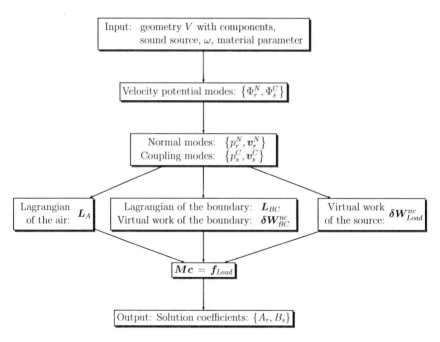

Figure 3.2: Flowchart of the algorithm for the coupling method.

3.2 Normal and Coupling Modes as Basis Functions

The normal and the coupling modes are the basis functions of the coupling method presented in Section 3.1. In this section we focus on the calculation of the velocity potential normal and coupling modes with Component Mode Synthesis (CMS) methods. A general review of CMS methods can be found in [25, 26, 23, 24]. These methods have been developed for the modelling of coupled substructures to improved the accuracy of the assembled model, which is of a reduced size. The reduction in size takes place at the component level in the modal space. In Chapter 4, we will describe numerical methods to compute these basis functions at the component level.

3.2.1 Normal Modes

The CMS is a technique to assemble models of several components. We describe the static and dynamic behaviour of each component in terms of a set of basis functions, the modes. These include normal modes and coupling modes, namely constraint or attachment modes. The reduction in size is achieved by truncating higher frequency modes at the component level. There exist different methods concerning the type of normal modes in the framework of CMS methods. The two most common CMS methods are the fixed-interface Craig-Bampton method [25] and the free-interface Craig-Chang method [23, 26]. In the following, we will derive the component modes of the different types. The vector matrix formulation of the Helmholtz equation inside an air component is

$$(\boldsymbol{K} - k^2 \boldsymbol{M})\boldsymbol{\Phi} = \boldsymbol{f}, \qquad (3.13)$$

where $\boldsymbol{\Phi}$ are the degrees of freedom (dof) of the velocity potential, $\boldsymbol{f}$ is the force, and $\boldsymbol{K}$ and $\boldsymbol{M}$ are the stiffness and the mass matrices, respectively. We separate the dofs into a set of interior dofs $\boldsymbol{\Phi}_I$ and a set of interface, or boundary dofs $\boldsymbol{\Phi}_B$. At the interface dofs there are two or more air components joined together. By using this partitioning, we write equation (3.13) as

$$\left(\begin{bmatrix} \boldsymbol{K}_{II} & \boldsymbol{K}_{IB} \\ \boldsymbol{K}_{IB}^T & \boldsymbol{K}_{BB} \end{bmatrix} - k^2 \begin{bmatrix} \boldsymbol{M}_{II} & \boldsymbol{M}_{IB} \\ \boldsymbol{M}_{IB}^T & \boldsymbol{M}_{BB} \end{bmatrix} \right) \begin{bmatrix} \boldsymbol{\Phi}_I \\ \boldsymbol{\Phi}_B \end{bmatrix} = \begin{bmatrix} \boldsymbol{f}_I \\ \boldsymbol{f}_B \end{bmatrix}, \qquad (3.14)$$

where $\boldsymbol{f}_B = \boldsymbol{0}$ if the boundary at the interface is free and $\boldsymbol{\Phi}_B = \boldsymbol{0}$ if the boundary at the interface is fixed. There are no forces acting on the interior of the air component ($\boldsymbol{f}_I = \boldsymbol{0}$).

Definition 3.6. Free-interface normal modes

The free-interface normal modes satisfy the homogeneous Neumann boundary condition along the boundary and using (3.13) they are defined by

$$(K - \lambda_r^{fr} M)\phi_r^{fr} := 0, \qquad r = 1, 2, ..., n_{dof}. \tag{3.15}$$

The superscript fr stands for free-interface normal modes and n_{dof} is the number of global dofs.

We divide the normal modes into a set of modes $r = 1, ..., N_r$ for further computations and a complementary set of modes $d = N_r + 1, ..., n_{dof}$ with $N_r \ll n_{dofs}$, which will be not used for the coupling procedure later on. Consequently, we have a matrix of free-interface normal modes

$$\phi^{fr} = \begin{bmatrix} \phi_r^{fr} & \phi_d^{fr} \end{bmatrix}. \tag{3.16}$$

Similarly, we arrange the related eigenvalues λ_r on the diagonal of the eigenvalue matrix Λ^{fr} and we divide them into sets with subscripts r and d

$$\Lambda^{fr} = \begin{bmatrix} \Lambda_r^{fr} & 0 \\ 0 & \Lambda_d^{fr} \end{bmatrix}. \tag{3.17}$$

If a component is unconstrained, then the normal mode set will contain rigid body modes with zero-valued eigenvalues.

Definition 3.7. Fixed-interface normal modes

The fixed-interface normal modes of a component are the eigenmodes of the component with the interface dofs fixed ($\Phi_B = 0$). With the partitioning the size of the eigenvalue problem is therefore reduced by the number of interface dofs. The related eigenvalue problem is governed by the elements of the mass and stiffness matrices associated with the interior dofs (subscript I) only and is defined by

$$(K_{II} - \lambda_r^{fi} M_{II})\phi_{I_r}^{fi} := 0, \qquad r = 1, 2, ..., n_{dof} - N_B, \tag{3.18}$$

where λ_r^{fi} are the fixed-interface eigenvalues. The superscript fi stands for fixed-interface normal modes.

The eigenvectors $\phi_{I_r}^{fi}$ compose the columns of the normal mode matrix ϕ^{fi}. We divide these modes into two sets with subscripts r and d as before. The normal mode matrix follows as

$$\phi^{fi} = \begin{bmatrix} \phi_r^{fi} & \phi_d^{fi} \end{bmatrix} = \begin{bmatrix} \phi_{I_r}^{fi} & \phi_{I_d}^{fi} \\ 0_{B_r} & 0_{B_d} \end{bmatrix}, \tag{3.19}$$

where 0_B relates to the dofs of the fixed boundary. There are no rigid body modes in equation (3.19) if the set of fixed boundary dofs is sufficient to constrain all rigid body modes of the unconstrained component.

For the coupling method in the frequency domain described in Section 3.1.2 we apply the free-interface CMS method to compute the velocity potential normal modes Φ^N. With (3.2) and (3.4) we yield the following properties. The pressure normal modes p_r^N are equal to free-interface modes, because they are proportional to the velocity potential normal modes due to the Helmholtz equation for the velocity potential

$$p_r^N = -\frac{\rho_0 c_0^2}{\omega} \Delta \Phi_r^N = \frac{\rho_0 c_0^2}{\omega} k_r^2 \Phi_r^N . \tag{3.20}$$

The velocity normal modes v_r^N with its component perpendicular to the boundary correspond to fixed-interface modes along the whole boundary of the air component related to equation (2.34) derived in Section 2.3.

For a rectangular two dimensional geometry V with dimensions L_x and L_y in x- and y-direction the pressure normal modes p_r^N have a cosine pattern as shown in Section 2.5. For the x-component of the velocity normal modes v_r^N there is a sinusoidal pattern in x-direction and a cosine pattern in y-direction. We present the x-component of the first velocity normal modes in Figure 3.3.

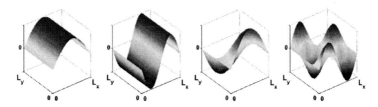

Figure 3.3: Fixed-interface normal modes of the velocity component in x-direction for the geometry $V = [0, L_x] \times [0, L_y]$.

3.2.2 Properties of Eigenmodes

It is worth mentioning that there exist important properties of the normal modes out of the free-interface as well as fixed-interface CMS methods [81]. These properties have influence on the implementation of the coupling method in Section 3.1.3.

Let us consider the variational formulation for the homogeneous Helmholtz equation

$$\int_V \nabla \phi \nabla \nu \, d\boldsymbol{x} - k^2 \int_V \phi \nu \, d\boldsymbol{x} = 0 , \quad \phi, \nu \in H^1(V) . \tag{3.21}$$

We note that this formulation involves two symmetric bilinear forms. The eigenvalues $\lambda = k^2$ are positive. This follows out of taking $\nu = \phi$, where λ becomes the quotient of two positive values.

Theorem 3.3. Orthogonality of the eigenmodes
We consider two solutions $(\lambda_\alpha, \phi_\alpha)$ *and* $(\lambda_\beta, \phi_\beta)$ *with* $\lambda_\alpha \neq \lambda_\beta$. *Then we yield the orthogonality property*

$$\int_V \phi_\alpha \phi_\beta \, d\boldsymbol{x} = 0,$$

$$\int_V \nabla\phi_\alpha \nabla\phi_\beta \, d\boldsymbol{x} = 0. \tag{3.22}$$

Proof:
This orthogonality property holds by applying the variational formulation (3.21) to ϕ_α ($\phi = \phi_\alpha$) and ϕ_β ($\nu = \phi_\beta$). Then we obtain

$$\int_V \nabla\phi_\alpha \nabla\phi_\beta \, d\boldsymbol{x} = \lambda_\alpha \int_V \phi_\alpha \phi_\beta \, d\boldsymbol{x},$$

$$\int_V \nabla\phi_\beta \nabla\phi_\alpha \, d\boldsymbol{x} = \lambda_\beta \int_V \phi_\beta \phi_\alpha \, d\boldsymbol{x}. \tag{3.23}$$

We subtract both equations in (3.23), use the symmetry of the bilinear forms, and obtain at $(\lambda_\alpha - \lambda_\beta) \int_V \phi_\alpha \phi_\beta \, d\boldsymbol{x} = 0$. Consequently, we conclude $\int_V \phi_\alpha \phi_\beta \, d\boldsymbol{x} = 0$, because of $\lambda_\alpha \neq \lambda_\beta$. Referring to (3.23) $\int_V \nabla\phi_\alpha \nabla\phi_\beta \, d\boldsymbol{x} = 0$ holds.

A special case of the orthogonality property (3.22) is

$$\int_V \phi_\alpha \, d\boldsymbol{x} = 0, \qquad \lambda_\alpha > 0. \tag{3.24}$$

To avoid such a constant mode ($\lambda = 0$, $\phi = $ constant) it is possible to use the additional constraint (3.24) in numerical methods to eliminate this solution without changing the other eigenmodes.

With the orthogonality property (3.22) we yield the orthogonality for the pressure normal modes as well as for the velocity normal modes. In our coupling method we neglect the constant eigenmode for the eigenvalue $\lambda = 0$.

For all eigensolutions $(\lambda_\alpha, \phi_\alpha)$, we denote by $\boldsymbol{v}_\alpha = \nabla\phi_\alpha$ the corresponding velocity of the fluid and by $p_\alpha = -\frac{\rho_0 c_0^2}{i\omega_\alpha}\Delta\phi_\alpha = \frac{\rho_0 \omega_\alpha}{i}\phi_\alpha$ the corresponding pressure. To become aware of the relation to the mode definition in equation (3.4), we remark that $\boldsymbol{v}_\alpha^* = i\omega_\alpha \boldsymbol{u}_\alpha^*$ holds for the time harmonic assumption, where $\boldsymbol{u}_\alpha^*$ is the instantaneous displacement. From the variational formulation (3.21) and with $\nu = \phi_\alpha$ we get

$$\int\limits_V |\nabla\phi_\alpha|^2 \, d\boldsymbol{x} - \frac{\omega_\alpha^2}{c_0^2} \int\limits_V \phi_\alpha^2 \, d\boldsymbol{x} = 0. \tag{3.25}$$

We show that this relation results from the conservation of total mechanical energy of the fluid during harmonic oscillation, that means the sum of the kinetic and the potential energies. For the time harmonic assumption we know that the instantaneous velocity is $\boldsymbol{v}_\alpha^*(\boldsymbol{x}, t) = \boldsymbol{v}_\alpha(\boldsymbol{x})e^{i\omega_\alpha t}$. We obtain the kinetic energy T_A inside the acoustic fluid to be

$$T_A = \frac{1}{2} \int\limits_V \rho_0 \left|\boldsymbol{v}_\alpha(\boldsymbol{x})\right|^2 \, d\boldsymbol{x} = \frac{\rho_0}{2} \int\limits_V \left|\nabla\phi_\alpha(\boldsymbol{x})\right|^2 \, d\boldsymbol{x}. \tag{3.26}$$

Using $p_\alpha^*(\boldsymbol{x}, t) = p_\alpha(\boldsymbol{x})e^{i\omega_\alpha t}$, the instantaneous displacement $\boldsymbol{u}_\alpha^*(\boldsymbol{x}, t) = \boldsymbol{u}_\alpha(\boldsymbol{x})e^{i\omega_\alpha t}$ and the relation between pressure and velocity (3.5) the potential energy U_A is given by

$$U_A = \frac{1}{2} \int\limits_V p_\alpha^*(\boldsymbol{x}, t)\nabla^T \boldsymbol{u}_\alpha^*(\boldsymbol{x}, t) \, d\boldsymbol{x} = -\frac{\rho_0 c_0^2}{2} \int\limits_V \left|\nabla^T \boldsymbol{u}_\alpha^*(\boldsymbol{x}, t)\right|^2 \, d\boldsymbol{x}$$

$$\tag{3.27}$$

$$= -\frac{1}{2\rho_0 c_0^2} \int\limits_V p_\alpha^2(\boldsymbol{x}) \, d\boldsymbol{x} = -\frac{\rho_0 \omega_\alpha^2}{2c_0^2} \int\limits_V \phi_\alpha^2(\boldsymbol{x}) \, d\boldsymbol{x}.$$

With (3.25) and the time harmonic assumption we conclude $T_A + U_A = 0$, which is equivalent to the conservation of total energy.

For the matrix vector formulation of the eigenvalue problem the stiffness matrix $\boldsymbol{K}$ and the mass matrix $\boldsymbol{M}$ arise from the variational formulation (3.21)

$$(\boldsymbol{K} - \lambda_r \boldsymbol{M})\boldsymbol{\phi}_r = 0, \qquad r = 1, 2, ..., n_{dof}. \tag{3.28}$$

The eigenvalue problem for the homogeneous system (3.13), which we have to solve for all kinds of normal modes, is self-adjoint. The kernel of the stiffness matrix $\boldsymbol{K}$ is of dimension 1 and the rank is $n_{dof} - 1$, if the total number of dofs is n_{dof}. We note that this holds, because $\boldsymbol{K}$ resulting with the variational formulation out of $\int_V \nabla\phi\nabla\nu \, d\boldsymbol{x}$ is zero if and only if $\phi = $ constant . The mass matrix $\boldsymbol{M}$ has blocks of zeros, if we force zero Dirichlet boundary conditions at the interface in the case of fixed-interface CMS methods. In this case we reduce the system by the number of dofs belonging to the related fixed interface. In both cases we yield $\lambda = k^2 = 0$ as an eigenvalue. We can also write the equation (3.28) as

$$\boldsymbol{\phi}_r^T(\boldsymbol{K} - \lambda_r \boldsymbol{M}) = 0, \qquad r = 1, 2, ..., \tag{3.29}$$

because the right eigenvector coincides with the left eigenvector. The orthogonality conditions for mass-normalised eigenvectors are

$$\phi_r^T M \phi_i = \delta_{ri},$$

$$\phi_r^T K \phi_i = \lambda_r \delta_{ri}, \tag{3.30}$$

where δ is the Kronecker Delta. The same holds for the normal modes of the velocity v^N. Because of this the matrix of the Lagrangian function of the air L_A contains many zero components. The block submatrix of the matrix L_A belonging to the normal mode components is a diagonal matrix. This is important for an efficient programming of the subroutine to compute L_A.

3.2.3 Coupling Modes

Related to the type of CMS method there exist also different types of coupling modes [23, 25, 26]. In contrast to the normal modes we force the coupling modes to have a specific pattern along a part of the boundary $\Gamma_C \subset \delta V$ and we only use the stiffness matrix K ($k = 0$ in equation (3.13)), which is called quasi-static. For the fixed-interface CMS the coupling modes are the constraint modes.

Definition 3.8. Constraint modes
The constraint modes have an inhomogeneous Dirichlet boundary condition along the boundary $\Gamma_C \subset \delta V$ and with (3.13) they are defined by

$$K^{red}\Phi_s^c := f^{red}, \quad s = 1, ..., N_s, \quad \Phi_s^c(x) := \begin{cases} g_s(x) & \forall x \in \Gamma_C \\ 0 & \forall x \in \delta V \backslash \Gamma_C \end{cases}. \tag{3.31}$$

The superscript c denotes the constraint modes.

Here, the system matrix K is reduced to K^{red} due to the prescribed dofs along Γ_C. For the free-interface CMS the related coupling modes are the attachment modes. In this case a function g_s is imposed along the interface boundary as an inhomogeneous Neumann boundary condition.

Definition 3.9. Attachment modes
The attachment modes have an inhomogeneous Neumann boundary condition along the boundary $\Gamma_C \subset \delta V$ and with (3.13) they are defined by

$$K\Phi_s^a := f, \quad s = 1, ..., N_s, \quad \frac{\partial \Phi_s^a(x)}{\partial n} := \begin{cases} g_s(x) & \forall x \in \Gamma_C \\ 0 & \forall x \in \delta V \backslash \Gamma_C \end{cases}. \tag{3.32}$$

The superscript a denotes the attachment modes and the superscript red stands for the reduced system related to the Dirichlet boundary condition.

For every attachment mode Φ_s^a the function g_s for the normal derivative is imposed along the coupling interface Γ_C. This results in a prescribed inhomogeneous Neumann boundary condition for the pressure constraint modes p_s^c. The size of the system matrix K remains $n_{dof} \times n_{dof}$. For the attachment modes as well as for the constraint modes, we have to solve the inhomogeneous Helmholtz equation once for every coupling mode.

Remark:

In the following, we will focus on the fried-interface CMS with constraint modes as coupling modes. In general, there are different types of constraint modes (also attachment modes). They are divided into nodal or modal and into static or non-static constraint modes. For a CMS method with nodal modes the number of nodal constraint modes is equal to the number of physical dofs along the interface (nodes of finite elements). For each of these dofs there exists exactly one constraint mode, which has the value one at this interface dof and is zero at all remaining interface dofs. If the CMS has modal constraint modes, then the number of modes can be much smaller then the number of dofs at the interface. Each modal constraint mode provides a certain function along the interface, a sine pattern, for instance. On the one hand, static constraint modes consider only the stiffness matrix for solving the inhomogeneous Helmholtz equation ($k = 0$). On the other hand, dynamic constraint modes ($k \neq 0$) are able to include inertial effects of the non-interface dofs. They have been developed in [98] amongst others as quasi-static modes demanding one centering frequency $\omega_{cent} = c_0 k_{cent}$ for all constraint modes.

Due to the mode definition in equation (3.4) the computed constraint modes for the velocity potential yield in attachment modes for the velocity coupling modes. For the former example of the two dimensional rectangular geometry we assume the interface to a coupling component to be at $x = 0$. We present the first three modal static constraint modes for the velocity in x-direction related to the coupling interface along $x = 0$ in Figure 3.4.

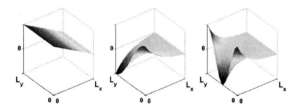

Figure 3.4: Modal static coupling modes of the velocity component in x-direction with a pattern $\cos(\frac{0\pi}{L_y})$, $\cos(\frac{1\pi}{L_y})$ and $\cos(\frac{2\pi}{L_y})$ related to the interface along $x = 0$.

We observe the characteristics of static coupling modes, the higher the sinusoidal argument $\frac{n\pi}{L_y}$ the faster the decrease of absolute amplitudes from the interface to the interior of the volume V. This is known as the near field character.

3.3 Boundary Models

A physical realistic boundary condition is reflecting and absorbing waves at the same time. Such boundaries are called absorbers [33, 34, 37, 79]. This characteristic is considered by its related frequency dependent impedance $Z(\omega)$, which we will describe in Appendix B. We note, that the coupling method does not need a direct discretization of the volume of the absorber. This is an advantage, because we do not need additional dofs to consider the absorber. We compute the needed absorber related impedance in a pre-process. The detailed calculation of the impedance is published by Buchschmid in [13, 14, 16]. In this section, we will focus on the model of single absorber components (Figure 3.5). We will describe absorber models as infinite ones first and discuss finite absorber models afterwards. At the end, we describe the numerical realization of the finite absorber component at the boundary of one air component (Figure 3.6) within the coupling method.

| Absorber | | Air Component | — | Absorber |

Figure 3.5: Single absorber component as a representative in the acoustic network.

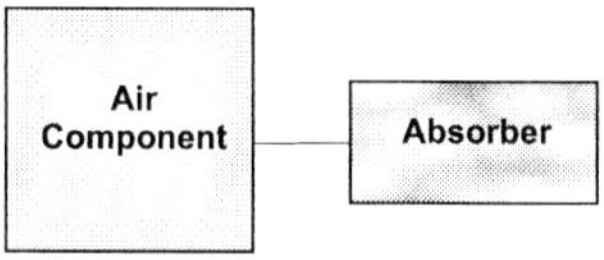

Figure 3.6: Absorber component in relation to a single air component.

3.3.1 Infinite Absorber Models

Absorbers are built up of different layers like air, plate-like structures and absorptive structures. They are categorized in passive absorbers (Figure 3.7), plate- and Helmholtz absorbers (Figure 3.8). Passive absorbers consist of porous materials like foams, mineral wool or cellular glass. The sound waves enter the pores of the absorber and initiate a vibration of the air in the interconnected pores. The kinetic energy of the sound field is reduced due to the flow resistance inside the porous absorber. Resonating plate-like absorbers are excited to vibrations by sound waves. Energy is reduced due to internal damping, which can be increased by absorption in a porous layer, installed in the air cushion behind the vibrating plate as in Figure 3.8 for example. In the following, we describe the model of such single layer or compound (multi layer) absorbers with the linear Theory of Porous Media (TPM) by de Boer [30]. Thereby, we model each layer with a constant thickness by differential equations.

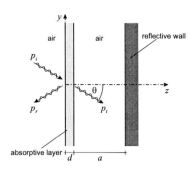

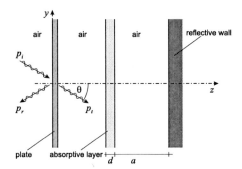

Figure 3.7: Passive absorber.   Figure 3.8: Plate-like absorber.

We describe the air layer for the pressure p with the homogeneous wave equation

$$\Delta p(\boldsymbol{x},t) - \frac{1}{c_0^2}\frac{\partial^2 p(\boldsymbol{x},t)}{\partial t^2} = 0. \tag{3.33}$$

We consider the plate-like layer to be out of homogeneous, linear-elastic, isotropic material and model it by the Lamé equation

$$\left(\lambda^H + \mu^H\right)\nabla\cdot\nabla^T\boldsymbol{u}_H(\boldsymbol{x},t) + \mu^H\Delta\boldsymbol{u}_H(\boldsymbol{x},t) = \rho^H\frac{\partial^2\boldsymbol{u}_H(\boldsymbol{x},t)}{\partial t^2}. \tag{3.34}$$

Here, $\boldsymbol{u}$ is the displacement and λ^H and μ^H are the Lamé constants. The superscript H denotes the homogeneous material in order to distinguish it from the constituents of the porous medium. The absorptive layer consists of solid and gas (air) and we describe it as a control volume with statistically distributed pores. The porous solid is assumed to be much stiffer than the air. This results in a structural compressibility of the material, which we compose out of an incompressible porous solid and a compressible gas. The volume fractions concept used in [29] provides the identification of the individual constituents in the control volume. Volume fractions n^α refer the volume element dv^α of the constituent α (porous solid S and gas G) to the volume element dv of the mixture. In addition, they relate the real density $\rho^{\alpha R}$ of the constituent with its partial density ρ^α. We define the balance of momentum for the solid phase considering the stresses by

$$-n^S\nabla p(\boldsymbol{x},t) + \left(\tilde{\lambda}^S + \mu^S\right)\nabla\left(\nabla^T\cdot\boldsymbol{u}_S(\boldsymbol{x},t)\right) +$$

$$\mu^S\Delta\boldsymbol{u}_S(\boldsymbol{x},t) + S_G\left(\boldsymbol{v}_G(\boldsymbol{x},t) - \boldsymbol{v}_S(\boldsymbol{x},t)\right) = \rho^S\boldsymbol{a}_S(\boldsymbol{x},t). \tag{3.35}$$

The balance of momentum for the gas phase ends up in

$$-n^G\nabla p(\boldsymbol{x},t) - S_G\left(\boldsymbol{v}_G(\boldsymbol{x},t) - \boldsymbol{v}_S(\boldsymbol{x},t)\right) = \rho^G\boldsymbol{a}_G(\boldsymbol{x},t). \tag{3.36}$$

Obeying the saturation condition $\sum_\alpha n^\alpha = 1$ and carrying out a linearization we result in the balance of mass

$$\frac{n^G}{R\,T}\frac{\partial p(\boldsymbol{x},t)}{\partial t} + \rho_0^{GR}\,n^G\nabla^T\cdot\boldsymbol{v}_G(\boldsymbol{x},t) + \rho_0^{GR}\,n^S\nabla^T\cdot\boldsymbol{v}_S(\boldsymbol{x},t) = 0\,. \tag{3.37}$$

ρ_0^{GR} is the constant factor of a Taylor series expansion. We simplify the equations (3.34), (3.35), (3.36), and (3.37) by expressing the displacement field in terms of a scalar potential and a vector potential. In this case all partial differential equations (3.34), (3.33), and systems of partial differential equations (3.35), (3.36) and (3.37) are transformed from the spatial domain to the wavenumber domain by means of the Fourier transform (see Appendix A.2). Consequently, they yield in ordinary differential equations. Afterwards we apply the Fourier transform from the time domain to the frequency domain due to the harmonic excitation. This concept allows different layers in one direction, but in the remaining two directions no variation inside the present layers is possible. We solve the resulting linear ordinary differential equations of second order with constant coefficients using an exponential approach. In the case of a compound absorber boundary conditions between two layers finally enable to solve for the unknowns. Apparently, only a few unknowns occur inside the system for each wavenumber and frequency dependent impedance of the infinite assumed absorber, which can be solved very efficiently. More details about the computation of a wavenumber and frequency dependent impedance for a single layer or compound absorber and the application to the coupling method in Section 3.1 is published by Buchschmid and Pospiech in [15, 91].

3.3.2 Finite Absorber Models

We restrict ourselves on a rectangular absorber, which is placed at the boundary of the three dimensional geometry V. For simplicity we describe the absorber with y- and z-coordinates. In general we have to transform the rectangular absorber to this plane. To involve the wavenumber and frequency dependent impedance we approximate the velocity pattern $v_{plate}(y,z,t)$ and the displacement pattern $u_{plate}(y,z,t)$ of the rectangular plate-like finite absorber with the eigenmodes of the clamped plate [88].

$$
\begin{aligned}
v_{plate}(y,z,t) &\approx \sum_{s=1}^{s_{max}}\sum_{t=1}^{t_{max}} v_{plate_{st}}(y,z)\left(B_{st}e^{i\omega t} - \overline{B_{st}}e^{-i\omega t}\right), \\[2mm]
u_{plate}(y,z,t) &\approx \sum_{s=1}^{s_{max}}\sum_{t=1}^{t_{max}} v_{plate_{st}}(y,z)\left(\frac{B_{st}}{i\omega}e^{i\omega t} + \frac{\overline{B_{st}}}{i\omega}e^{-i\omega t}\right), \\[2mm]
v_{plate_{st}}(y,z) &= b_s(y)\Theta(y)\Theta(L_y - y)b_t(z)\Theta(z)\Theta(L_z - z),
\end{aligned}
\tag{3.38}
$$

with

$$\Theta(y) = \begin{cases} 0, & y \le 0 \\ 1, & y > 0 \end{cases} \tag{3.39}$$

$$b_s(y) = sin\left(\frac{\beta_s y}{L_y}\right) - sinh\left(\frac{\beta_s y}{L_y}\right) + \gamma_s\left(cosh\left(\frac{\beta_s y}{L_y}\right) - cos\left(\frac{\beta_s y}{L_y}\right)\right)$$

$$b_t(z) = sin\left(\frac{\beta_s z}{L_z}\right) - sinh\left(\frac{\beta_t z}{L_z}\right) + \gamma_s\left(cosh\left(\frac{\beta_t z}{L_z}\right) - cos\left(\frac{\beta_t z}{L_z}\right)\right)$$

$$\gamma_s = \frac{sinh(\beta_s) - sin(\beta_s)}{cosh(\beta_s) - cos(\beta_s)}$$

$$\beta_1 = 4.73004074$$

$$\beta_2 = 7.85320462$$

$$\beta_3 = 10.99560783$$

$$\beta_s = (s + 0.5)\pi, \quad s > 3.$$

b_s and b_t denote the eigenmodes of the clamped beam. They are zero at both endpoints, they have zero first derivatives at both endpoints, and they are pairwise orthogonal to each other. Θ is the Heaviside step function. Θ filters the pattern with the size of the plate absorber $[0, L_y] \times [0, L_z]$ out of the infinite plane in y- and z-direction.

We choose the time-dependent terms of $v_{plate}(y, z, t)$ in equation (3.38) equal to the time-dependent terms of the velocity coupling modes in equation (3.1) (see Section 3.1.2). We approximate each plate mode $v_{plate_{st}}(y, z)$, belonging to one coupling mode, with a finite complex Fourier series

$$v_{plate_{st}}(y, z) \approx \sum_{m=-m_{max}}^{m_{max}} \sum_{n=-n_{max}}^{n_{max}} E_{m_s n_t} e^{i(k_y(m)y + k_z(n)z)}. \tag{3.40}$$

We compute the Fourier coefficients by

$$E_{m_s n_t} = \frac{1}{L_y^{rep} L_z^{rep}} \int\limits_0^{L_z^{rep}} \int\limits_0^{L_y^{rep}} v_{plate_{st}}(y, z) e^{-i(k_y(m)y + k_z(n)z)} dy\, dz \tag{3.41}$$

with

$$k_y(m) = \frac{2\pi m}{L_y^{rep}}, \quad k_z(n) = \frac{2\pi n}{L_z^{rep}}.$$

The definition of $L_y^{rep} > L_y$ and $L_z^{rep} > L_z$ is discussed in [13] in detail depending on the choise of the trial functions for each plate mode.

We calculate the Lagrangian for the plate absorber with the kinetic and the potential energies T_{TPM} and U_{TPM} using [87] by

$$T_{TPM}(t) = \frac{1}{2} \int\limits_0^{L_z} \int\limits_0^{L_y} \rho d \left| v_{plate}(y, z, t) \right|^2 dy\, dz, \tag{3.42}$$

where ρ is the constant density and d the constant thickness of the absorber. The potential energy of the absorber writes

$$U_{TPM}(t) \;=\; \frac{1}{2} \cdot \frac{Ed^3}{12(1-\nu^2)} \int_0^{L_z} \int_0^{L_y} \left| \frac{\partial^2 u_{plate}(y,z,t)}{\partial y^2} + \frac{\partial^2 u_{plate}(y,z,t)}{\partial z^2} \right|^2 + \qquad (3.43)$$

$$2(1-\nu) \left(\frac{\partial^2 u_{plate}(y,z,t)}{\partial y^2} \frac{\partial^2 u_{plate}(y,z,t)}{\partial z^2} - \frac{\partial^2 u_{plate}(y,z,t)}{\partial y \partial z} \right) dy\,dz \,,$$

where E denotes the Young's modulus and ν the Poisson's ratio of the absorber. Introducing (3.38) and (3.41) in (3.42) and (3.43) and using the Parseval's Theorem [8] we yield the Lagrangian of the finite absorber

$$\int_0^T L_{BC}(t)\,dt \;=\; \int_0^T T_{BC}(t) \;-\; U_{BC}(t)\,dt \;\approx \qquad (3.44)$$

$$\frac{1}{2} \sum_{s=1}^{s_{max}} \sum_{t=1}^{t_{max}} 2 B_{st}\overline{B_{st}} T L_y^{rep} L_z^{rep} \sum_{m=-m_{max}}^{m_{max}} \sum_{n=-n_{max}}^{n_{max}} \rho d \left| E_{m_s n_t} \right|^2 \;-$$

$$\frac{1}{2} \sum_{s=1}^{s_{max}} \sum_{t=1}^{t_{max}} \frac{2 B_{st}\overline{B_{st}}}{\omega^2} T L_y^{rep} L_z^{rep} \sum_{m=-m_{max}}^{m_{max}} \sum_{n=-n_{max}}^{n_{max}} \left(\frac{(k_y(m)^2 + k_z(n)^2)^2 Ed^3}{12(1-\nu^2)} \right) \left| E_{m_s n_t} \right|^2 \;.$$

Summarizing (3.44) we get

$$\int_0^T L_{BC}(t)\,dt \;=\; \frac{T}{\omega} L_y^{rep} L_z^{rep} \sum_{s=1}^{s_{max}} \sum_{t=1}^{t_{max}} B_{st}\overline{B_{st}} \sum_{m=-m_{max}}^{m_{max}} \sum_{n=-n_{max}}^{n_{max}} Im(Z(m,n,\omega)) \left| E_{m_s n_t} \right|^2 \quad (3.45)$$

with

$$Im(Z(m,n,\omega)) = \left(\rho d\omega - \frac{(k_y(m)^2 + k_z(n)^2)^2 Ed^3}{12(1-\nu^2)\omega} \right) \,.$$

The term $Im(Z(m,n,\omega))$ stands for the imaginary part of the plate impedance $Z(m,n,\omega)$ [27], which is wavenumber and frequency dependent. We apply the same approach for the virtual work of the non-conservative damping forces and get

$$\int_0^T \delta W_{BC}^{nc} dt \;=\; - \int_0^T \int_0^{L_z} \int_0^{L_y} C\, v_{plate}(y,z,t)\, \delta u_{plate}(y,z,t)\, dy\,dz\,dt \,. \qquad (3.46)$$

Including (3.38) and (3.40) yields in

$$\int_0^T \delta W_{BC}^{nc} dt = \tag{3.47}$$

$$-\frac{T}{i\omega} L_y^{rep} L_z^{rep} \sum_{s=1}^{s_{max}} \sum_{t=1}^{t_{max}} \left(B_{st} \overline{\delta B_{st}} - \overline{B_{st}} \delta B_{st} \right) \sum_{m=-m_{max}}^{m_{max}} \sum_{n=-n_{max}}^{n_{max}} Re(Z(m,n,\omega)) |E_{m_s n_t}|^2$$

Here, $Re(Z(m,n,\omega)) = C$ stands for the real part of the plate impedance and represents the damping of the absorber.

Remark:

The coefficients B_{st} and its complex conjugated value $\overline{B^{st}}$ represent the ingoing and the outgoing wave. Both, the Lagrangian in equation (3.45) and the virtual work of the nonconservative damping forces in equation (3.47), include mixed terms of the coefficients B^{st}, $\overline{B^{st}}$, δB_{st}, and $\overline{\delta B_{st}}$. This is necessary to compose the matrices L_{BC} and δW_{BC}^{nc} as described in Section 3.1.3.

3.4 Sound Source Models

Sound sources are treated as source terms in the governing equations (see Sections 2.2 and 2.3). In the inhomogeneous Helmholtz equation (2.28), the source terms f and q for (external) force per unit volume and volume velocity per unit volume, respectively, are already considered. Non-linear mechanisms related to Lighthill's acoustic analogy [66] can be incorporated with the right-hand side as additional source term, which are beyond the scope of this work. In general we distinguish between sound sources and reflectors. Sound sources are moving under a force like loudspeakers. Reflectors reflect, scatter and diffract sound waves in the sound field. Examples are a window inside a room, a picture fixed at the wall or a carpet lying on the floor.

In the following we discuss basic concepts of sound sources and how to incorporate them in the framework of our coupling method in Section 3.1. In the acoustic network analogy we have to model sound sources (Figure 3.9) and their relation to the air component (Figure 3.10), where they are located (see Figure 3.1).

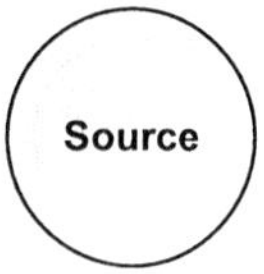

Figure 3.9: Single sound source out of the acoustic network.

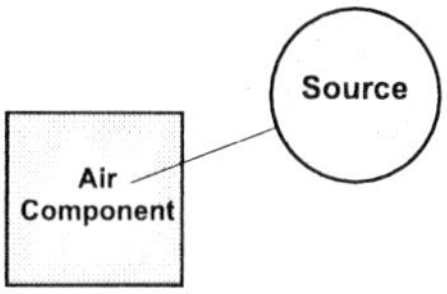

Figure 3.10: Sound source in relation to an air component.

First elementary source model

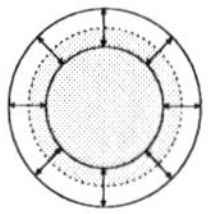

A point source has no spatial dimension and is modelled as a monopole. At the point source the sound field is singular and can be represented by the Dirac delta function $\delta(\boldsymbol{x})$

$$\int_{-\infty}^{\infty} \phi(\boldsymbol{x})\delta(\boldsymbol{x} - \boldsymbol{x}_0)d\boldsymbol{x} = \phi(\boldsymbol{x}_0). \tag{3.48}$$

Here, ϕ denotes an arbitrary continuous function and $\boldsymbol{x}_0$ denotes the location of the point source. The Dirac delta function belongs to generalized functions, see [67] for more information. In an unbounded domain considering no further sources the solution for a harmonic point monopole at $\boldsymbol{x}_0$ is the Green function $g(\boldsymbol{x}, \boldsymbol{x}_0)$ with a singularity at the point $\boldsymbol{x}_0$

$$g(\boldsymbol{x}, \boldsymbol{x}_0) = \frac{e^{-ikr}}{4\pi r}, \qquad r = |\boldsymbol{x} - \boldsymbol{x}_0| . \qquad (3.49)$$

Note that $g(\boldsymbol{x}, \boldsymbol{x}_0) = g(\boldsymbol{x}_0, \boldsymbol{x})$, which is so-called acoustic reciprocity. That means the solution is invariant in exchanging source and receiver. This is valid for all linear acoustic systems and concedes technical and economic benefits in the experimental determination of vibroacoustical transmission paths. Consequently quieter products in industry can be processed. Considering the volume velocity $q(r,t) = Q_0 e^{i(\omega t - kr)}$ in equation (2.18), the concentrated rate of change of rate of mass introduction at $\boldsymbol{x}_0$ is represented by

$$i\omega \rho_0 q(\boldsymbol{x}, \omega) = Q_0 \delta(\boldsymbol{x} - \boldsymbol{x}_0), \qquad (3.50)$$

where Q_0 is the total source strength. The solution for the sound pressure in the frequency domain using (3.49) and (3.50) yields

$$p(r) = \frac{i\omega \rho_0 Q_0}{4\pi r} e^{-ikr} . \qquad (3.51)$$

Obviously, the absolute value of the sound pressure is decreasing by increasing the distance r to the monopole source and is dependent on the wavenumber.

The volume velocity of a point monopole in a closed cavity does net work against the resistive component of a local pressure, which it induces. In general, we can express the pressure as the sum of all the modal pressures at the source point, see equation (3.1). Because of the discontinuity at the position of the monopole, the number of needed modes tends to infinity as the distance to the source location decreases to zero. Therefore, we recommend to assume a finite but small sound source (referred to as circular or spherical source in the following). The modal series solution in equation (3.1) will then converge acceptably quickly.

Second elementary source model

Vibrating panels produce displacements of the fluid, therefore they are acting as boundary forces and do work on the fluid. They belong to category 2, discussed in Section 2.4. Boundary forces f on an inviscid fluid act purely in the direction normal to the local surface. We can describe them by the Heaviside function and $\nabla^T f = f\delta(\boldsymbol{x} - \boldsymbol{x}_1) - f\delta(\boldsymbol{x} - \boldsymbol{x}_2)$. Second elementary sources are two virtual point sources of equal magnitude Q_0 and opposite sign, a typical dipole. A rigid sphere undergoing transverse oscillations generates a sound field close to a point dipole. Any rigid body with at least one cross-sectional dimensional transverse to the oscillating axis much smaller than an acoustic wavelength generates a dipole-like sound field. These sources do not produce net volume displacement, but create fluid momentum fluctuations (see the source term in the law of momentum conservation (2.7)). We derive the pressure distribution for a dipole by

$$p(r) = \frac{i\omega\rho_0 Q_0}{4\pi} \left(\frac{e^{-ikr_2}}{r_2} - \frac{e^{-ikr_1}}{r_1} \right). \tag{3.52}$$

Here, r_1 and r_2 denote the distance to the first and second monopole, respectively. The distance between the two sources is smaller than the acoustic wavelength. A dipole source does not radiate sound in all directions equally like the monopole. The directivity pattern looks like a figure-8. There are two regions where sound is radiated very well, and two regions where sound cancels. In Table 3.2 both elementary sound models are summarized.

Table 3.2: Summary of the two main sound sources.

Sound source	Solution for the pressure
First elementary sound model	$p(r) = \frac{i\omega\rho_0 Q_0}{4\pi r} e^{-ikr}$
Second elementary sound model	$p(r) = \frac{i\omega\rho_0 Q_0}{4\pi} \left(\frac{e^{-ikr_2}}{r_2} - \frac{e^{-ikr_1}}{r_1} \right)$

Numerical implementation

For the numerical implementation of elementary sound sources we will consider monopoles and circular sources as sound sources on the one hand. On the other hand, we will use finite line sources and investigate the resulting sound field. With both, circular and line sources, we will excite waves to analyse the effects on the sound distribution inside the air volume of interest. With such sinusoidal patterns we can also simulate a dipole, the second elementary sound source. We include an elementary dipole by forcing an elliptical sound source boundary with a sinusoidal pattern with wavenumber $\frac{2\pi x}{l_S}$. l_S denotes the length of the elliptic shaped sources boundary. All sound sources with boundaries have in common that we have to apply the force rectangular to their boundary. These forces are acting on the velocity normal and coupling modes $\boldsymbol{v}^N$ and $\boldsymbol{v}^C$, respectively, and build the right-hand side $\boldsymbol{f}_{Load}$ of the linear system of equations (3.12) for the coupling coefficients A_r and B_s. In Chapter 5, we will discuss the numerical results for a monopole as well as for a circular sound source and line sources in acoustic benchmark geometries.

4 Numerical Methods for Computing the Basis Functions

In the previous chapter we discussed the coupling method for room acoustical problems with realistic boundaries in detail. Therein we needed basis functions, the normal modes and the coupling modes, to approximate the sound field in a closed room, like a cavity. The normal modes depend on the geometry of the acoustic volume and the coupling modes depend on the position and the type of boundary condition. In the present section we analyse appropriate numerical methods to compute the basis functions as a pre-process before the computation process of the coupling method. The basis functions are smooth and arbitrarily often differentiable. The Spectral Method (SM) and the Spectral Finite Element Method (SEM) amongst others are very suitable for such kind of functions. In the following we focus on the Spectral Method and then on the Spectral Finite Element Method. We will present their principle of operation, their convergence properties and we will compare both methods for simple examples of basis function computations. At the end of this chapter we focus on the problem of smooth derivatives out of Finite Element solutions.

4.1 Spectral Method

The SM [19, 42, 106] is a collocation method [56, 63, 102] to solve partial differential equations such as the Helmholtz equation (2.28) on simple symmetric domains, like cuboids, spheres or cylinders. The idea of the SM is to approximate a global derivative by using a collocation approach. This method shows spectral convergence for smooth problems as it is the case for the basis functions of the coupling method in the frequency domain (Chapter 3). In the following we introduce the SM for 1d problems.

4.1.1 Approach for One Dimensional Domains

The bounded domain $V = [-1, 1]$ is subdivided by supporting points $x_j = -1 + hj$ for $j = 1, ..., N$ with $h = \frac{2}{N}$. The interpolation polynomial u is equal to the solution p of interest at the supporting points

$$u(x_j) = p(x_j) := p_j \, . \tag{4.1}$$

The Fourier transform of p is denoted by $\hat{p}$. The discrete Fourier coefficients are given by

$$\hat{p}_k = h \sum_{j=1}^{N} e^{-ikx_j} p_j, \qquad \text{supp } \hat{p} \subset \left[-\frac{\pi}{h}, \frac{\pi}{h} \right] \, . \tag{4.2}$$

For more details about the Fourier transform and the discrete Fourier transform see Appendix A.2. We obtain the interpolation polynomial u by the superposition of exponentials multiplied with the related discrete Fourier coefficients

$$u(x) = \frac{1}{2} \sum_{k=-\frac{N}{2}+1}^{\frac{N}{2}} e^{ikx} \hat{p}_k \, . \tag{4.3}$$

Hence, we express the interpolation polynomial at every supporting point with the help of exponentials and discrete Fourier coefficients of p:

$$u(x_j) = \frac{1}{2} \left(\sum_{k=-\frac{N}{2}+1}^{\frac{N}{2}-1} e^{ikx_j} \hat{p}_k + \frac{1}{2} \left(e^{-i\frac{N}{2}x_j} \hat{p}_{-\frac{N}{2}} + e^{i\frac{N}{2}x_j} \hat{p}_{\frac{N}{2}} \right) \right), \qquad \hat{p}_{-\frac{N}{2}} = \hat{p}_{\frac{N}{2}} \, . \tag{4.4}$$

Theorem 4.1. Differentiation matrix
The differentiation matrix $D_N \in \mathbb{R}^{N \times N}$ for $V = [-1, 1]$ and the supporting points $x_j = -1 + hj$, $j = 1, ..., N$ is given by

$$(D_N)_{ij} := \begin{cases} \dfrac{\frac{\pi}{2h} \cos\left(\frac{\pi x}{h}\right) \sin(x) - \frac{1}{2} \sin\left(\frac{\pi x}{h}\right)}{\frac{2\pi}{h} \sin^2\left(\frac{x}{2}\right)}, & i \neq j, \\[4mm] 0, & i = j, \end{cases} \qquad i, j = 1, ..., N \, . \tag{4.5}$$

Proof:
If u_δ is the interpolation polynomial of the Dirac delta function δ (see equation (3.48)) with its Fourier transform $\hat{\delta} = h$, then u_δ has the form

$$\begin{aligned} u_\delta(x) &= \frac{h}{2} \left(\sum_{k=-\frac{N}{2}+1}^{\frac{N}{2}-1} e^{ikx} + \frac{1}{2} \left(e^{-i\frac{N}{2}x} + e^{i\frac{N}{2}x} \right) \right) \\[3mm] &= \frac{h \cot\left(\frac{x}{2}\right) \sin\left(\frac{\pi x}{h}\right)}{2\pi} = \underbrace{\frac{\sin\left(\frac{\pi x}{h}\right)}{\frac{2\pi}{h} \tan\left(\frac{x}{2}\right)}}_{=: S_h^*(x)} \, . \end{aligned} \tag{4.6}$$

With that we represent the single function values p_j as a superposition of the previous interpolation polynomial S_h^*

$$u(x) = \sum_{j=1}^{N} S_h^*(x - x_j) p_j. \tag{4.7}$$

The differentiation matrix $\boldsymbol{D}_N$ results from (4.7) as

$$(\boldsymbol{D}_N)_{ij} = S_h^{*'}(x_i - x_j) = S_h^{*'}((i-j)h)$$

and with calculating $S_h^{*'}$ follows equation (4.5).

Clustered grids

To avoid instabilities near the boundary, which is a serious issue for rational polynomials [106], we work with clustered grids. That means the supporting points are more dense at the boundary of the interval. In the following, we use the Chebyshev nodes in $V = [-1, 1]$

$$x_j = cos\left(\frac{\pi j}{N}\right), \qquad j = 0, 1, ..., N. \tag{4.8}$$

We apply the barycentric Lagrange interpolation through the Chebyshev nodes to arrive at the more stable spectral collocation with related weights λ_j

$$u(x) = \prod_{j=0}^{N}(x - x_j) \sum_{j=0}^{N} \frac{p(x_j)\lambda_j}{(x-x_j)}, \quad \lambda_j = \frac{1}{w'_{N+1}(x_j)} = \begin{cases} \frac{(-1)^j}{2}, & j \in \{0, N\} \\ (-1)^j, & j = 1, ..., N-1 \end{cases}. \tag{4.9}$$

The differentiation matrix $\boldsymbol{D}_N$ follows straight on. The second derivative yields out of $\boldsymbol{D}_N^{(2)} = (\boldsymbol{D}_N)^2$.

Numerical treatment of boundary conditions

For a boundary problem such as the Helmholtz equation, the treatment of boundary conditions is inevitable. We consider the homogeneous Robin boundary condition, which is a combination of the well known Dirichlet and Neumann boundary condition. The Robin boundary condition is equal to the acoustic impedance boundary condition as defined in Definition 2.2. We apply the boundary condition left and right of the reference domain $V = [-1, 1]$

$$\alpha \frac{\partial p(x)}{\partial n} + \beta p(x) = 0, \quad x = -1, \quad \alpha, \beta \in \mathbb{C},$$
$$\gamma \frac{\partial p(x)}{\partial n} + \delta p(x) = 0, \quad x = 1, \quad \gamma, \delta \in \mathbb{C}. \tag{4.10}$$

For the strong formulation of the inhomogeneous Helmholtz equation $\Delta p(x) + k^2 p(x) = f(x)$ on V the matrix vector equation is

$$\left(\begin{bmatrix} \gamma \boldsymbol{D}_N(1,:) \\ \boldsymbol{D}_N^2(2:N,:) \\ \alpha \boldsymbol{D}_N(N+1,:) \end{bmatrix} + \begin{bmatrix} \delta \\ k^2 \mathbb{1}_{N-1} \\ \beta \end{bmatrix} \right) \boldsymbol{p} = \begin{pmatrix} 0 \\ \boldsymbol{f}(2:N) \\ 0 \end{pmatrix}. \quad (4.11)$$

$\mathbb{1}_{N-1}$ stands for the identity matrix with dimensions $(N-1) \times (N-1)$ and k for the wavenumber. Note that we have to take care of the order of the nodes (see (4.8)). For more than one spatial direction we have to consider all points along the boundary to include such combinations of boundary conditions.

Tensor product grids in multi dimensions

For problems in more than one dimension we set up a tensor product grid based on Chebyshev points independently in each direction. The easiest way to solve a problem on a tensor product grid is to use tensor products. In linear algebra they are also known as Kronecker products. The Kronecker product of two matrices $\boldsymbol{A}$ and $\boldsymbol{B}$ is denoted by $\boldsymbol{A} \otimes \boldsymbol{B}$. The order of matrices in Kronecker products, such as in the Kronecker products for differentiation matrices in multi dimensions, differs depending on the underlying programming language. In MATLAB [76] the order of some chosen examples follows as

$$\begin{aligned}
\text{2d:} \quad \Delta &= \mathbb{1}_{N+1} \otimes \boldsymbol{D}_N^{(2)} + \boldsymbol{D}_N^{(2)} \otimes \mathbb{1}_{N+1}, \\
\tfrac{\partial^2}{\partial x \partial y} &= \boldsymbol{D}_N \otimes \boldsymbol{D}_N, \\
\end{aligned}$$

$$\quad (4.12)$$

$$\begin{aligned}
\text{3d:} \quad \Delta &= \mathbb{1}_{N+1} \otimes \mathbb{1}_{N+1} \otimes \boldsymbol{D}_N^{(2)} + \mathbb{1}_{N+1} \otimes \boldsymbol{D}_N^{(2)} \otimes \mathbb{1}_{N+1} + \\
& \quad \mathbb{1}_{N+1} \otimes \mathbb{1}_{N+1} \otimes \boldsymbol{D}_N^{(2)}, \\
\tfrac{\partial^3}{\partial z^3} \tfrac{\partial}{\partial y} \tfrac{\partial^2}{\partial x^2} &= \boldsymbol{D}_N^{(3)} \otimes \boldsymbol{D}_N \otimes \boldsymbol{D}_N^{(2)}.
\end{aligned}$$

This tensor product grids have the advantage that the concept of one dimension is reused in multi dimensions. The drawback is that only simple symmetric domains can be described by tensor grids.

4.1.2 Convergence

Theorem 4.2. Accuracy of Fourier spectral differentiation
Let $p \in L^2(\mathbb{R})$ have a μth derivative ($\mu \geq 1$) of bounded variation and let w be the μth spectral derivative of p on the grid $h\mathbb{Z}$. Then the following estimates hold for all supporting points $x \in h\mathbb{Z}$

i) if p has $m - 1$ continuous derivatives in $L^2(\mathbb{R})$ for some $m \geq \mu + 1$ and a mth derivative of bounded variation, then

$$\left| p^{(\mu)}(x) - w(x) \right| \leq O(h^{m-\mu}) \quad \forall m \in \mathbb{N}, \tag{4.13}$$

ii) the interpolation error for an integrable $p \in C^\infty$ and its spectral interpolant u is

$$\left| p^{(\nu)}(x) - u^{(\nu)}(x) \right| \leq O(h^m) \quad \forall m \in \mathbb{N}, \tag{4.14}$$

where the superscript (ν) denotes the derivative of the related function.

iii) if p is analytical in addition, then the interpolation error

$$\left| p^{(\nu)}(x) - u^{(\nu)}(x) \right| = O(e^{-\frac{c}{h}}) \tag{4.15}$$

expresses the exponential convergence, also called spectral convergence.

Further details and the proof of Theorem 4.2 can be found in [106].

Summarizing this section, the properties of the SM yield that smooth functions have rapidly decaying Fourier coefficients. Hence the aliasing errors introduced by discretization are small. With usually smooth solutions in room acoustic problems on the one side we gain spectral convergence and we can easily include impedance-like boundary conditions. A remarking drawback is that the method is only applicable to simple and symmetric geometries related to the global derivative operators and the tensor product grids. In addition, we have an exponentially growing computational effort for higher dimensions and growing complexity due to the tensor product grids including global derivative operators. Consequently the method is only adaptive for simple symmetric geometries, but is easy to implement and efficient for computations.

4.2 Spectral Finite Element Method

To compute the solution for the Helmholtz equation for complex domains we will focus on Finite Element methods [20, 51, 75, 103, 110], especially on the Spectral Finite Element method (SEM), in this section. Finite Element methods for time harmonic acoustics have been an active research area for nearly 40 years. A special class of Finite Element methods are the hp-methods. These methods enable to refine the elements in space (h-refinement) and allow to increase the polynomial degree inside each element (p-refinement) in order to achieve exceptionally fast, thus exponential convergence rates. SEM belongs to this class. In the following, we describe the concept and the properties of the SEM [57, 78, 92].

4.2.1 Variational Formulation and Choice of Basis Expansion

Many numerical methods are based on the variational or weak formulations of boundary value problems. Let us consider the Helmholtz problem with Neumann boundary conditions and without sources on the bounded domain V

$$\Delta p(\boldsymbol{x}) + k^2 p(\boldsymbol{x}) = 0, \qquad \forall \boldsymbol{x} \in V \subset \mathbb{R}^3,$$

$$\frac{\partial p(\boldsymbol{x})}{\partial \boldsymbol{n}} = g(\boldsymbol{x}), \qquad \forall \boldsymbol{x} \in \Gamma_N = \delta V, \ \beta \in \mathbb{C}.$$

(4.16)

Multiplying with test functions u, integrating over the domain V and applying an integration by parts results in the weak formulation

$$\int_V \nabla p(\boldsymbol{x}) \nabla u(\boldsymbol{x}) - k^2 p(\boldsymbol{x}) u(\boldsymbol{x}) \, dV = \int_{\delta V} g(\boldsymbol{x}) u(\boldsymbol{x}) \, dS, \qquad \forall p, u \in H^1(V). \quad (4.17)$$

For Helmholtz problems given on a general bounded domain V, the natural trial and test spaces for the variational formulations are the Sobolev spaces H^k (see Appendix A.1). If a solution of the variational problem (4.17) exists, then the problem is weakly solvable and the solution $p \in H^1(V)$ is a variational or a weak solution of the boundary value problem (4.16). After the domain partition $V = \bigcup_{e=1}^{N_{el}} V^e$ we approximate p with trial functions out of the trial space, a finite dimensional subspace of H^1, which is spanned by the test functions for approximating u.

Theorem 4.3.
Let $H^\delta \subset H^1$ be any finite dimensional subspace. Given $p, u \in H^\delta$ and $g \in L^2(V)$ (4.17) has a unique solution and is equivalent to solving the following square matrix equation

$$\left(\boldsymbol{K} - k^2 \boldsymbol{M} \right) \hat{\boldsymbol{p}} = \boldsymbol{f}, \qquad \boldsymbol{K}, \boldsymbol{M} \in \mathbb{R}^{n_{dof} \times n_{dof}}, \ \hat{\boldsymbol{p}}, \boldsymbol{f} \in \mathbb{R}^{n_{dof} \times 1} \qquad (4.18)$$

$\boldsymbol{K}$ *denotes the global stiffness matrix and* $\boldsymbol{M}$ *denotes the global mass matrix. The right-hand side* $\boldsymbol{f}$ *incorporates the normal derivative of p at the domain boundaries and n_{dof} is the number of global degrees of freedom.*

Proof:

Let $\{\psi_i : 1 \leq i \leq n_{dof}\}$ be a basis of the finite dimensional subspace H^δ. There follows $\boldsymbol{K}_{ij} = \int\limits_V \nabla\psi_j(\boldsymbol{x})\nabla\psi_i(\boldsymbol{x})\,dx$, $\boldsymbol{M}_{ij} = \int\limits_V \psi_j(\boldsymbol{x})\psi_i(\boldsymbol{x})\,dx$, and $\boldsymbol{f}_i = \int\limits_V g(\boldsymbol{x})\psi_i(\boldsymbol{x})\,dx$. With [9, 54] amongst others results the proof of Theorem 4.3.

$\hat{\boldsymbol{p}}$ is the vector of the global coefficients of the basis functions. These coefficients are mapped to the vector of local coefficients

$$\hat{\boldsymbol{p}} = \boldsymbol{A}\hat{\boldsymbol{p}}_{loc} \tag{4.19}$$

with

$$\hat{\boldsymbol{p}} \in \mathbb{R}^{n_{dof} \times 1},\ \boldsymbol{A} \in \mathbb{R}^{n_{dof} \times N_{el}n_e},\ \hat{\boldsymbol{p}}_{loc} \in \mathbb{R}^{N_{el}n_e \times 1}\,.$$

$\boldsymbol{A}$ represents the mapping of the vector of local coefficients $\hat{\boldsymbol{p}}_{loc}$ to the vector of global coefficients $\hat{\boldsymbol{p}}$. N_{el} is the number of elements in the geometry and n_e is the number of degrees of freedom in each element. Finally, we write the approximated solution in V as a linear combination of trial functions, which are non-zero in their related element V^e and zero everywhere else in V

$$p^\delta(\boldsymbol{x}) = \sum_i \psi_i(\boldsymbol{x})\hat{\boldsymbol{p}}_{loc_i},\quad \forall \boldsymbol{x} \in V\,. \tag{4.20}$$

Here, the superscript δ stands for the hp-finite dimensional approximation of the pressure solution p.

Modal C^0 basis expansion

To satisfy $p^\delta \in H^1(V)$, the basis expansion has to be at least continuous almost everywhere. Nevertheless, the choice of the basis expansion has an important effect on the numerical conditioning. The SEM basis expansion consists of Jacobi polynomials [57].

Definition 4.1. Jacobi polynomials
Jacobi polynomials or hypergeometric polynomials are orthogonal polynomials of the following form

$$J_l^{\alpha,\beta}(x) = \frac{\Gamma(\alpha+l+1)}{n!\Gamma(\alpha+\beta+l+1)} + \sum_{m=0}^l \binom{l}{m}\frac{\Gamma(\alpha+\beta+l+m+1)}{\Gamma(\alpha+m+1)}\left(\frac{x-1}{2}\right)^m\,. \tag{4.21}$$

The superscripts have to satisfy $\alpha, \beta > 1$.

The Jacobi polynomials can also be constructed with the help of a three-term recursion [57]. An important property of these polynomials is the following orthogonal relation

$$\int\limits_{-1}^{+1}(1-x)^\alpha(1+x)^\beta J_l^{\alpha,\beta}(x)J_m^{\alpha,\beta}(x)\,dx = \begin{cases} 0, & l \neq m \\ C, & l = m \end{cases}\,. \tag{4.22}$$

Here, C depends on α, β, and l with

$$C = \frac{2^{\alpha+\beta+1}}{2l+\alpha+\beta+1} \frac{\Gamma(l+\alpha+1)\Gamma(l+\beta+1)}{l!\,\Gamma(l+\alpha+\beta+1)}.$$

A class of symmetric polynomials, the ultraspheric polynomials, refer to the choice $\alpha = \beta$. The Legendre polynomials ($\alpha = \beta = 0$) and the Chebyshev polynomials ($\alpha = \beta = -0.5$) are well-known representatives out of this class. In SEM we separate the basis expansions in nodal and modal ones. We will use the following continuous modal C^0 basis.

Definition 4.2. Modal C^0 basis
The modal p-type basis with polynomial degrees 1 to P on the reference interval $\xi \in [-1,1]$ is defined by

$$\psi_l(\xi) = \begin{cases} \frac{1-\xi}{2}, & l = 0, \\[2mm] \left(\frac{1-\xi}{2}\right)\left(\frac{1+\xi}{2}\right)J_{l-1}^{1,1}(\xi), & 0 < l < P, \\[2mm] \frac{1+\xi}{2}, & l = P. \end{cases} \tag{4.23}$$

$J_l^{\alpha,\beta}$ *denotes the Jacobi polynomial of equation (4.21) with polynomial degree l.*

In (4.23) ψ_0 and ψ_P correspond to the trial function of the one dimensional linear FEM. The remaining trial functions are the so-called interior modes or bubble modes. In general, we can define the shape of interior modes in (4.23) as any polynomial, which satisfies the zero value at the intervals boundary points. However, using the Jacobi polynomials $J_l^{1,1}$ in the following maintains a high degree of orthogonality and generates a local mass matrix whose interior coupling produces a penta-diagonal system. For the integrals contained in the components of the matrix vector formulation in (4.18), we apply the Gauss-Lobatto-Legendre quadrature with the quadrature points ξ_j and weights $w_j^{0,0}$

$$\xi_j = \begin{cases} -1, & j = 1, \\[2mm] \xi_{j-2,Q-2}^{1,1}, & j = 2,...,Q-1, \\[2mm] 1, & j = Q, \end{cases} \tag{4.24}$$

$$w_j^{0,0} = \frac{2}{Q(Q-1)\left(J_{j-2}^{0,0}(\xi_j)\right)^2}, \qquad j = 1,...,Q.$$

The quadrature points $\xi_{j-2,Q-2}^{1,1}$ are the $(Q-2)$ zeros of the Jacobi polynomial $J_{Q-2}^{1,1}$. They are clustered and symmetrically distributed in $V = [-1,1]$. Using the Gauss-Lobatto-Legendre quadrature of order Q, a polynomial with a degree up to $2Q - 3$ is integrated exactly. Consequently, we choose the number of quadrature points $Q = P + 2$ for integrating the trial functions up to the polynomial degree P.

The differentiation matrix $D \in \mathbb{R}^{Q \times Q}$ is related to the Gauss-Legendre quadrature points in (4.24). D is given by

$$
D_{ij} = \begin{cases}
-\dfrac{Q(Q-1)}{4}, & i = j = 1, \\[2ex]
\dfrac{J_{Q-1}^{1,1}(\xi_i)}{(\xi_i - \xi_j) J_{Q-1}^{1,1}(\xi_j)}, & i \neq j, \ 1 \leq i,j \leq Q, \\[2ex]
0, & 2 \leq i = j \leq Q - 1, \\[2ex]
\dfrac{Q(Q-1)}{4}, & i = j = Q.
\end{cases}
\tag{4.25}
$$

Tensor product expansion in multi dimensions

The majority of the bases used in hp-FE methods can be expressed in terms of a product of one-dimensional functions or tensor products. This is similar to the tensor product grids for the SEM in Section (4.1). Expansions constructed with this concept allow many numerical operations to be performed efficiently using the sum factorization or tensor product techniques. In three dimensions the related hexahedral tensor product expansion for the reference domain $V = [-1, 1]^3$ is

$$
\psi_{l_1}(\xi_1)\psi_{l_2}(\xi_2)\psi_{l_3}(\xi_3), \quad 0 \leq l_1, l_2, l_3 \leq P, \quad \xi_1, \xi_2, \xi_3 \in [-1, 1].
\tag{4.26}
$$

In two dimensions, for instance there exist four vertex modes, $4(P-1)$ edge modes and $(P-1)^2$ interior modes. The basis functions for $\alpha = \beta = 1$ in one and in two dimensions are presented in Figure 4.1 and 4.2 for a polynomial degree of $P = 3$ exemplarily.

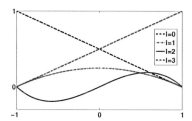

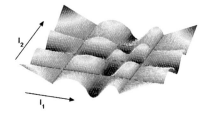

Figure 4.1: Legendre basis functions in one dimension with polynomial degree $P = 3$.

Figure 4.2: Legendre basis functions in two dimension with polynomial degree $P = 3$.

To construct a globally continuous C^0 basis from elemental or local contributions we need to introduce a local to global assembly process. This process is referred to as direct stiffness summation or global assembly. Thus for the numerical implementation, therefore, we have to consider the connectivity and the modal edge or face orientation [57] at the boundaries between adjacent elements within the mapping matrix $\boldsymbol{A}$ (see equation (4.19)). We have to guarantee that all vertex and boundary modes simply match the adjacent elements vertex and boundary modes of similar shape. When considering an expansion with more than one edge node (Figure 4.2), we need to take care of the local orientation of the element. Depending on the orientations of the local coordinate systems within the element, the sign of odd-ordered modes may need to be reversed. The reason for the sign negation is that the elemental modal shapes are defined with respect to the local coordinate system. If the local coordinate systems are orientated so that the two neighbouring coordinates are in opposite directions, then the sign of the odd-shaped modes needs to be reversed.

The use of a triangular or a tetrahedral SEM in computational fluid dynamics is relatively limited compared to quadrilateral and hexahedral SEM discretizations. This is due to the numerical efficiency of the algorithm, especially in the context of cost per time-step in time-dependent computations.

4.2.2 Convergence

To give a general error estimate for hp-type FE methods such as the SEM, we consider the one-dimensional Helmholtz problem with a solution $p \in H^s(V)$.

Theorem 4.4.
Assuming a discretization on a uniform mesh of equispaced elements of size h, the general error estimate in the energy norm $\| \cdot \|_E$ for the hp-type extension process writes

$$\|p - p^\delta\|_E = Ch^{\mu-1}P^{-(s-1)}\|p\|_s. \tag{4.27}$$

$\mu = \min(s, P+1)$ *and C is independent of h, P, and p, but depends on s. We also assume the present forces $f \in H^{s-2}(V)$, so that it does not appear in the error estimate.*

We remark that, if the solution p is smooth enough to have bounded derivatives for $s \geq P + 1$, then the error estimate in Theorem 4.4 shows that we achieve exponential convergence as we increase the polynomial order P (p-type extension). The proof for the previous theorem is stated in [4]. A similar estimate is valid for the Spectral Element version.

Theorem 4.5.
We assume the solution $p \in H_0^s(V)$ and the force $f \in H^\rho(V)$. Then the valid error estimate in the H^1 norm results in

$$\|p - p^\delta\|_1 = C \left(P^{1-s}\|p\|_s + P^{-\rho}\|f\|_\rho \right). \tag{4.28}$$

Here, the number of elements is constant and homogeneous Dirichlet boundary conditions are prescribed on δV.

The proof of Theorem 4.28 can be found in [71].

Related to a linear elliptic eigenvalue problem, such as the Helmholtz equation with a zero right-hand side (compare equation (2.28)), in FE methods the accuracy for the computed eigenmodes and corresponding eigenvalues deteriorates with the size of the eigenvalues. For one spatial dimension there exists a standard a priori error estimate for h-FE methods.

Theorem 4.6.
In one spatial dimension the absolute error of the computed eigenvalues λ_m^h, with $\lambda_m^h = (k_m^h)^2$, for h-FE methods with basis expansions in C^s can be estimated by

$$\|\lambda_m^h - \lambda_m\| \leq Ch^{2(P-s)}\lambda_m^{(P+1)/(s+1)}, \quad C \in \mathbb{R}.$$

Here, λ_m is the exact mth eigenvalue, where the eigenvalues are ordered by increasing magnitude. λ_m^h is the h-FE approximation for the mth eigenvalue, h is the size of the elements in V, C is a constant independent of h and λ_m and P is the polynomial degree of the basis functions.

This estimate corresponds to a Galerkin FE method in the standard self-adjoint case, where the eigenvalues are real and non-negative. Consequently, the absolute value is redundant due to the known inequality $\lambda_m \leq \lambda_m^h$. This originates from the Rayleigh quotient theorem. But the latter inequality only holds, if there is no so-called variational crime included such like reduced integration, use of incompatible modes or mass lumping. The last factor in the estimate of Theorem 4.6 shows, that the higher the eigenvalue becomes, the more difficult is it to compute it under a prescribed accuracy. The proof of Theorem 4.6 can be found in [103].

To compute the basis functions in the coupling method (Section 3.1), we need to compute eigenfunctions and related eigenfrequencies. The accuracy is decreasing with the increasing magnitude of the eigenfrequency. If M is the number of reasonable eigenfrequencies and N is the total number of degrees of freedom, then the common relation

$$M = rN, \quad 0 < r < 1, \tag{4.29}$$

holds, which is typical of all FE approximations. A rule of thumb is $r \approx 0.1$. The inverse strategy is, that we want to compute M eigenfunction eigenfrequency pairs with a certain accuracy ϵ and need to know the related h-discretization or the polynomial degree P for the domain V in d spatial dimensions. This gives us the total number of dofs.

Theorem 4.7.
In d spatial dimensions with an FE-basis in C^0 the number M of eigenfunction eigenvalue pairs with a maximal relative error ϵ of the computed eigenvalues $\lambda_1, ..., \lambda_M$ is given by

$$M = r_0 \epsilon^{\frac{d}{(2P)}} N.$$

N is the total number of degrees of freedom and P is the polynomial degree of the basis expansion in the FE method. r_0 is defined by

$$r_0 = \left| \frac{D(\gamma\beta U_d)^{\frac{2}{d}}}{4\pi^2} - 1 \right|^{-\frac{d}{(2P)}}.$$

Here, D depends on the type of element, γ is a constant depending on the shape of the element, β depends on the mesh topology and on the polynomial degree P with $\beta = N_{el}/N_{np}$. N_{el} is the number of elements, N_{np} is the total number of nodal points in the mesh, and U_d is the hyper-volume of the unit ball ($U_1 = 2$, $U_2 = \pi$, $U_3 = 4\pi/3$).

The Theorem is related to the physical problem of the modal density described in Section 2.5.1. The proof is similar to that in [41]. First, we apply the estimate in Theorem 4.6 to M and N

$$\left| \lambda_M^h - \lambda_M \right| \cong Ch^{2P}\lambda_M^{(P+1)}, \tag{4.30}$$

$$\left| \lambda_N^h - \lambda_N \right| \cong Ch^{2P}\lambda_N^{(P+1)}. \tag{4.31}$$

We want the relative errors of the significant eigenvalues to be smaller than the given tolerance ϵ, thus

$$\frac{\left| \lambda_m^h - \lambda_m \right|}{\lambda_m} \le \epsilon, \qquad m = 1, ..., M.$$

For the highest significant eigenvalue, $m = M$, the inequality turns into the equality

$$\frac{\left| \lambda_M^h - \lambda_M \right|}{\lambda_M} = \epsilon. \tag{4.32}$$

Dividing (4.30) by (4.31) and applying (4.32) yields

$$\epsilon \cong \left| \frac{\lambda_N^h}{\lambda_N} - 1 \right| \left(\frac{\lambda_M}{\lambda_N} \right)^P. \tag{4.33}$$

The Weyl asymptotic formula, extended by Courant [22, 90], for large eigenvalues of Helmholtz problems with homogeneous boundary conditions is

$$\lambda_m \approx Qm^{\frac{2}{d}}, \qquad Q = 4\pi^2 \left(\frac{1}{U_d vol(V)} \right)^{\frac{2}{d}}. \tag{4.34}$$

$vol(V)$ is the volume of the domain V. The formula (4.34) is remarkable, because the large eigenvalues depend only on the size of the domain V. They do not depend on the shape of V or on the parameters of the boundary conditions.

For the discrete spectral radius λ_N^{hP} one way is to use the theorem that bounds λ_N^h in the h-Galerkin version by the maximum eigenvalue of the eigenvalue problem [51]. For the standard h-FE methods with C^0 basis expansions and assuming a uniform mesh of well proportioned elements of size h leads to

$$\lambda_N^h \leq \frac{D}{h^2}, \tag{4.35}$$

where D depends on the type of element, for example $D = 24$ for $d = 2$, $P = 1$, and a consistent mass matrix [51].

Finally, we need a relation between the mesh parameter h and the total number of degrees of freedom N. We assume a uniform mesh with well-proportioned elements and we assume that the boundary effects affecting the h-N relation are negligible. It holds that

$$vol(V) = N_{el} vol(V^e), \tag{4.36}$$

where N_{el} is the total number of elements in the mesh, $vol(V)$ is the volume of the computational domain, and $vol(V^e)$ is the volume of the element. Furthermore

$$vol(V^e) = \gamma h^d, \tag{4.37}$$

where γ is a constant depending on the shape of the element. In addition,

$$N_{el} = \beta N_{nP}, \tag{4.38}$$

where N_{nP} is the total number of nodal points in the mesh and β depends on the mesh topology and on the polynomial degree P. $N_{nP} \cong N/N_{ndof}$, where N_{ndof} is the number of degrees of freedom per node. For the Helmholtz problem we assume $N_{ndof} = 1$. Combining (4.36)-(4.38) we obtain

$$h \cong \left(\frac{vol(V)}{\gamma \beta N} \right)^{\frac{1}{d}}. \tag{4.39}$$

By rearranging (4.33), (4.34), (4.35), and (4.39) we finally end up in the M-N relation of Theorem 4.7.

4.3 Comparison between the Spectral Method and the Spectral Finite Element Method

In the following we want to compare the SM (Section 4.1) and the SEM (Section 4.2). For the comparison we use a simple benchmark geometry $V = [-1, 1]$ in three dimensions on which we compute the eigenfunctions. These eigenfunctions are the basic step to the normal modes of the coupling method (Sections 3.1.2 and 3.2). For the eigenfunctions we prescribe homogeneous Neumann boundary conditions in both methods.

We are going to compute eigenfunctions, which equal to the velocity potential normal modes Φ^N. For these modes we have to enforce the homogeneous Neumann boundary condition at all boundaries of V. This boundary condition resembles to an acoustical hard surface, which is totally reflecting, as described in equation (2.36). The eigenfunctions and the related eigenfrequencies are computed from the Helmholtz equation, see equation (2.28). Both, the eigenfunctions Φ_r^N as well as the eigenfrequencies ω_r, are necessary for the computation of the pressure normal modes p_r^N and the velocity normal modes v_r^N as defined in equation (3.4). The analytical solution for the eigenfrequencies ω_r in V results from the help of the analytical solution for eigenfrequencies of totally reflecting general cuboids (see equation (2.43))

$$\omega_r := \omega_{r_1 r_2 r_3} = c_0 k_{r_1 r_2 r_3} = c_0 \sqrt{\left(\frac{r_1 \pi}{2}\right)^2 + \left(\frac{r_2 \pi}{2}\right)^2 + \left(\frac{r_3 \pi}{2}\right)^2}, \qquad r_1, r_2, r_3 \in \mathbb{Z}. \quad (4.40)$$

Implementation of the boundary condition

For the SM, as described in Section 4.1, we take the original form of the three dimensional Helmholtz equation without source terms in the frequency domain (compare to equation (2.28))

$$\Delta\Phi(\boldsymbol{x}) + k^2 \Phi(\boldsymbol{x}) = 0, \qquad \forall \boldsymbol{x} \in V. \quad (4.41)$$

With $N + 1$ Chebyshev supporting points, defined in equation (4.8), in every coordinate direction we write the Laplace operator such as in (4.12)

$$\boldsymbol{\Delta} = \mathbb{1}_{N+1} \otimes \mathbb{1}_{N+1} \otimes \boldsymbol{D}_N^{(2)} + \mathbb{1}_{N+1} \otimes \boldsymbol{D}_N^{(2)} \otimes \mathbb{1}_{N+1} + \boldsymbol{D}_N^{(2)} \otimes \mathbb{1}_{N+1} \otimes \mathbb{1}_{N+1}.$$

To account for the homogeneous Neumann boundary condition at all boundaries of V

$$\frac{\partial \Phi(\boldsymbol{x})}{\partial \boldsymbol{n}} = 0, \qquad \forall \boldsymbol{x} \in \delta V,$$

we compute the first derivative in every coordinate direction with respect to the underlying tensor grid points

$$\boldsymbol{D}_x = \mathbb{1}_{N+1} \otimes \mathbb{1}_{N+1} \otimes \boldsymbol{D}_N,$$

$$\boldsymbol{D}_y = \mathbb{1}_{N+1} \otimes \boldsymbol{D}_N \otimes \mathbb{1}_{N+1},$$

$$\boldsymbol{D}_z = \boldsymbol{D}_N \otimes \mathbb{1}_{N+1} \otimes \mathbb{1}_{N+1}.$$

To apply the boundary condition, we have to find the row numbers related to the boundary points, which conform to the eight vertices (b_{xyz}), to the six faces (b_{xy}, b_{xz}, b_{yz}), with two faces rectangular to each coordinate direction, and to the twelve edges (b_x, b_y, b_z), with four edges each being parallel to one coordinate direction. These vectors of row numbers have distinct entries. That means the vectors of faces have no entries related to grid points at edges or vertices and the vectors of edges have no entries related to grid points at vertices. Finally, we overwrite the rows in the matrix $\mathbf{\Delta}$ due to the needed related first derivative expressions to include the corresponding Neumann boundary condition

$$\mathbf{\Delta}(\boldsymbol{b}_{xyz},:) \;=\; \boldsymbol{D}_x(\boldsymbol{b}_{xyz},:) + \boldsymbol{D}_y(\boldsymbol{b}_{xyz},:) + \boldsymbol{D}_z(\boldsymbol{b}_{xyz},:),$$

$$\mathbf{\Delta}(\boldsymbol{b}_{xy},:) \;=\; \boldsymbol{D}_z(\boldsymbol{b}_{xy},:),$$

$$\mathbf{\Delta}(\boldsymbol{b}_{xz},:) \;=\; \boldsymbol{D}_y(\boldsymbol{b}_{xz},:),$$

$$\mathbf{\Delta}(\boldsymbol{b}_{yz},:) \;=\; \boldsymbol{D}_x(\boldsymbol{b}_{yz},:), \qquad (4.42)$$

$$\mathbf{\Delta}(\boldsymbol{b}_x,:) \;=\; \boldsymbol{D}_y(\boldsymbol{b}_x,:) + \boldsymbol{D}_z(\boldsymbol{b}_x,:),$$

$$\mathbf{\Delta}(\boldsymbol{b}_y,:) \;=\; \boldsymbol{D}_x(\boldsymbol{b}_y,:) + \boldsymbol{D}_z(\boldsymbol{b}_y,:),$$

$$\mathbf{\Delta}(\boldsymbol{b}_z,:) \;=\; \boldsymbol{D}_x(\boldsymbol{b}_z,:) + \boldsymbol{D}_y(\boldsymbol{b}_z,:) .$$

To not perturb the homogeneous Neumann boundary condition, we have to delete the corresponding boundary entries in the identity operator with

$$\boldsymbol{I}_{1d} \;=\; \mathbb{1}_{N+1}, \qquad \boldsymbol{I}_{1d}(1,1) := 0, \quad \boldsymbol{I}_{1d}(N+1,N+1) := 0,$$
$$\boldsymbol{I}_{3d} \;=\; \boldsymbol{I}_{1d} \otimes \boldsymbol{I}_{1d} \otimes \boldsymbol{I}_{1d} . \qquad (4.43)$$

With the boundary condition adapted matrices for the Laplace operator (4.42) and with the identity operator (4.43) the discretized form of the Helmholtz equation (4.41) is

$$\left(\mathbf{\Delta} \,+\, k^2 \boldsymbol{I}_{3d}\right) \mathbf{\Phi} \;=\; \mathbf{0} . \qquad (4.44)$$

We solve this general eigenvalue problem with the standard eigenvalue solver *eig* in MAT-LAB [76], considering the matrix $\mathbf{\Delta}$ being not symmetric and $\boldsymbol{I}_{3d}$ being not regular.

For the SEM, described in Section 4.2, we take the discretized matrix vector equation of the variational formulation of the Helmholtz equation for the velocity potential Φ as stated in equation (4.18)

$$\left(K - k^2 M\right)\hat{\Phi} = f.$$

Here, the right-hand side f involves all normal derivatives of the wanted solution Φ. If we enforce the homogeneous Neumann boundary condition, we yield $f = 0$ and end up in the general eigenvalue problem

$$\left(K - k^2 M\right)\hat{\Phi} = 0, \tag{4.45}$$

where the stiffness matrix K is symmetric and the mass matrix M is symmetric and positive definite, as in every discrete Galerkin FE formulation. To solve (4.45) we take the *eig* solver in MATLAB.

Comparison

For the comparison of the resulting eigenfrequencies we compute the eigenfrequencies with the SM using an equally growing N in all three coordinate directions. In addition, we compute the eigenfrequencies with the SEM using one element in V and an equally growing polynomial degree P in all three coordinate directions. For both methods the global numbers of degrees of freedom, which equals to the size of the global matrices in the general eigenvalue problems (4.44) and (4.45), are $(N + 1)^3$ and $(P + 1)^3$, respectively. Then, we compute the relative error due to the analytical eigenfrequencies $\omega_{r_1 r_2 r_3}$ with equation (4.40). In Figure 4.3 the relative errors of the eigenfrequencies ω_{100} and ω_{211} are plotted for both methods in relation to the global number of degrees of freedom.

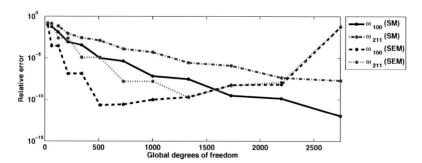

Figure 4.3: Comparison of the relative errors of the eigenfrequencies ω_{100} and ω_{211} between the SM and the SEM (with one element in V) for totally reflecting boundaries in V to the number of global degrees of freedom.

The comparison shows, that up to 1331 degrees of freedom the relative error of the second eigenfrequency is better for the SEM than for the SM. For the 20th eigenfrequency the relative error for the SEM is better than for the SM up to the global number of 2197 degrees of freedom. We observe the relative error of the SM is decreasing monotonically with a spectral convergence rate for both eigenfrequencies ω_{100} and ω_{211} as stated in Theorem 4.2. In contrast to the SM, the relative error of frequencies computed by the SEM with one element in V is getting worse with an increasing number of global degrees of freedom. This is related to the growing condition of the global mass matrix M shown in Figure 4.5, which is caused by the increasing polynomial degree P of the basis functions in the SEM (see equation (4.23)). In a second step, we compute the eigenfrequencies ω_{100} and ω_{211} with the SEM for V subdivided into eight equal cube elements. In Figure 4.4 we compare the new relative errors of the SEM with the former ones of the SM.

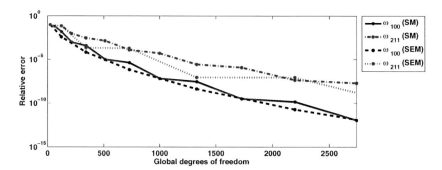

Figure 4.4: Comparison of the relative error in the eigenfrequencies ω_{100} and ω_{211} between the SM and the SEM, now with with eight cube elements in V, for totally reflecting boundaries in V to the number of global degrees of freedom.

The relative error for both eigenfrequencies ω_{100} and ω_{211} computed with the SEM is decreasing monotonically and for all numbers of global degrees of freedom the error is less or equal to the related relative error by the SM. In Figure 4.3 as well as in Figure 4.4 we would get similar results for all other eigenfrequencies.

The lack of increase of the relative error for an increasing polynomial degree P is due to the condition of the global mass matrix M for eight equal cube elements in V. The comparison of the condition of M in the SM and the SEM is depicted in Figure 4.5.

Due to the h-refinement for a fixed number of global degrees of freedom the polynomial degree for eight cube elements is always less than for one element in V. Consequently the condition of the global mass matrix M for eight elements is always less than the condition for the global mass matrix for one element if the number of global degrees of freedom is fixed. For a higher polynomial degree P the condition of the global mass matrix for eight cube elements would increase similar to the condition of the global mass matrix for one element with a comparable high polynomial degree P.

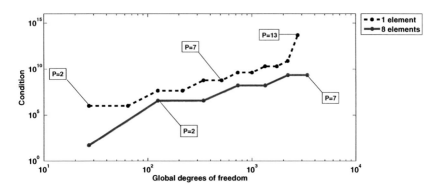

Figure 4.5: Comparison of the condition number of the global mass matrix M of the SEM with one element in V and of the global mass matrix M of the SEM with eight cube elements in V.

Result:

The relative error of the eigenfrequencies approximated by the SM is decreasing monotonically with a spectral convergence rate. The relative error of the eigenfrequencies computed with the SEM decreases even faster compared to the SM if we use an appropriate h-refinement. But it also can increase again, which is due to the polynomial degree P of the SEM basis resulting in a worse condition of the mass matrix M. Another way out of the increasing relative error is an adequate preconditioning of the global SEM mass matrix M.

4.4 Strategy for Smooth Derivatives

In Section 4.2 and 4.3 we proposed the SEM for computing the velocity potential normal and coupling modes $\{\Phi^N, \Phi^C\}$ as basis functions for the coupling method described in Section 3.1. To gain the velocity normal and coupling modes and the pressure coupling modes we need the derivatives of the velocity potential described in equation (3.4)

$$v_r^N(x) = i\nabla\Phi_r^N(x), \quad v_s^C(x) = i\nabla\Phi_s^C(x), \quad p_s^C(x) = -\frac{\rho_0 c_0^2}{\omega}\Delta\Phi_s^C(x).$$

Using the SEM on a mesh with rectangular elements or a mesh on a rectangular domain V, there occurs no problem in computing the gradients of the velocity potential Φ. But if we have a non-rectangular geometry, there occur high slopes and oscillations near the element boundaries with increasing amplitudes for higher polynomial degrees P of the SEM basis. This oscillations are due to the C^0 continuity of the basis functions in the SEM. This phenomenon is inherent for all FE methods with C^0 basis expansions. Nevertheless, there are many problems, for instance the strain-elasticity problem in structural dynamics, where the derivatives of the primal variables are of interest. Consequently, in the last decades there have been many approaches to yield in a satisfying solution for computing smooth derivatives of the primal variables.

A wide spread approach is the mixed formulation. Here both variables of interest are computed at the same time with different integration schemes and supporting points in every element [11]. However this kind of method is not suitable for our kind of problem, because if we have a skew boundary, where the normal derivative of the pressure is zero for all eigenmodes, then we cannot prescribe the boundary condition for the velocity in x-, y- and z-direction independently.

Another alternative are C^1 Finite Elements, described for example in [1, 86], which include for example Hermite basis functions. But these approaches are restricted to rectangular geometries V or cannot be expanded to higher polynomial degrees for the basis functions. In addition, extensive post-processing averaging techniques have been developed in the FE frame. For example in [92] an averaging for a nodal SEM is described, which is restricted to $P = 2$.

Another approach are global methods such as the global kernel method [55, 65, 69, 99], which uses global basis functions with arbitrary continuity on a meshless domain. Non-uniform rational B-splines [89, 52, 53] in contrast are global basis functions on a mesh with an arbitrary boundary. There is a growing interest in this approach, because the method offers great flexibility and precision for handling both analytical and freeform shapes.

4.4.1 Patch Recovery Technique

In the following we describe the patch recovery technique to yield the gradients of the velocity potential modes. The idea of the technique is based on [68], but we will not construct the gradient at the element nodes. We will use the given quadrature points and we will include the given boundary information to the computation of the gradients on the boundary of the volume. We derive the patch recovery technique in two spatial dimensions to yield smooth non-oscillating gradients of the velocity potential normal modes $\nabla \Phi^N$. We compute the velocity potential normal modes by computing the eigenmodes including homogeneous Neumann boundary conditions with the matrix vector equation (4.18) as described in Section 4.2

$$(K + k_j^2 M)\hat{\Phi}_j = 0, \qquad j = 1, ..., dof_{glob}. \tag{4.46}$$

dof_{glob} is the number of global degrees of freedom, which is equal to the size of the system matrices, the stiffness matrix K and the mass matrix M. Let us assume Φ to be one velocity potential eigenmode of equation (4.46) with its related eigenfrequency $\omega = c_0 k$ to describe the patch recovery technique.

Step 1: Mesh refinement

Because of the SEM and the tensor grid we have Q^2 quadrature points in each quadrangle element (see (4.24)) of the discretized mesh $V = \bigcup_{n=1}^{N_{el}} V^n$. With the help of these points we get a new mesh with $(Q-1)^2$ times as much elements as before $V = \bigcup_{n=1}^{(Q-1)^2 N_{el}} \tilde{V}^n$. This mesh refinement is sketched in Figure 4.6 for one single quadrangle element of the original mesh and for $Q = 4$, examplarily.

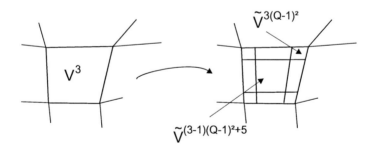

Figure 4.6: Mesh refinement from the originally mesh $V = \bigcup_{e=1}^{N_{el}} V^e$ to the finer mesh $V = \bigcup_{e=1}^{N_{el}(Q-1)^2} \tilde{V}^e$, $Q = 4$.

Step 2: Biquadratic interpolation for centre node gradients

We apply a biquadratic interpolation in every quadrangle element $\tilde{V}^n$

$$I_\Phi(x,y) := a_1 + a_2 x + a_3 y + a_4 x^2 + a_5 xy + a_6 y^2 + a_7 x^2 y + a_8 xy^2 \,. \tag{4.47}$$

To get the coefficients $\boldsymbol{a} = (a_1, ..., a_8)^T$ we solve the linear equation system

$$\boldsymbol{X}\boldsymbol{a} = \boldsymbol{f}, \tag{4.48}$$

with

$$\boldsymbol{X} = \begin{bmatrix} 1 & x_A & y_A & x_A^2 & x_A y_A & y_A^2 & x_A^2 y_A & x_A y_A^2 \\ 1 & x_B & y_B & x_B^2 & x_B y_B & y_B^2 & x_B^2 y_B & x_B y_B^2 \\ 1 & x_C & y_C & x_C^2 & x_C y_C & y_C^2 & x_C^2 y_C & x_C y_C^2 \\ 1 & x_D & y_D & x_D^2 & x_D y_D & y_D^2 & x_D^2 y_D & x_D y_D^2 \\ 0 & 0 & 0 & 2 & 0 & 2 & 2y_A & 2x_A \\ 0 & 0 & 0 & 2 & 0 & 2 & 2y_B & 2x_B \\ 0 & 0 & 0 & 2 & 0 & 2 & 2y_C & 2x_C \\ 0 & 0 & 0 & 2 & 0 & 2 & 2y_D & 2x_D \end{bmatrix}, \quad \boldsymbol{f} = \begin{pmatrix} \Phi(x_A, y_A) \\ \Phi(x_B, y_B) \\ \Phi(x_C, y_C) \\ \Phi(x_D, y_D) \\ -k^2\Phi(x_A, y_A) \\ -k^2\Phi(x_B, y_B) \\ -k^2\Phi(x_C, y_C) \\ -k^2\Phi(x_D, y_D) \end{pmatrix}.$$

I_Φ interpolates the velocity potential Φ and we can approximate Φ_x and Φ_y at the centre point (x_n, y_n) (see Figure 4.7) using $\boldsymbol{a}$ by

$$\begin{pmatrix} \Phi_x(x_n, y_n) \\ \Phi_y(x_n, y_n) \end{pmatrix} \approx \nabla I_\Phi(x_n, y_n) = \begin{pmatrix} a_2 + 2a_4 x_n + a_5 y_n + 2a_7 x_n y_n + a_8 y_n^2 \\ a_2 + a_5 x_n + 2a_6 y_n + a_7 x_n^2 + 2a_8 x_n y_n \end{pmatrix}. \tag{4.49}$$

We apply this biquadratic interpolation to all quadrangles $\tilde{V}^e \subset V$, $e = 1, ..., (Q-1)^2 N_{el}$.

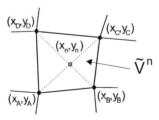

Figure 4.7: For the gradient at the centre point (x_n, y_n) in the quadrangle element $\tilde{V}^n$ all related element nodes add information with the help of the eigenmode solution of the Helmholtz equation (4.46).

Step 3: Bilinear interpolation for interior node gradients

After approximating the gradient at the centre points we take all interior nodes (x_i, y_i) of the mesh and compute the related gradients using a bilinear interpolation in the patch around each interior node

$$
\begin{aligned}
I_{\Phi_x}(x,y) &:= a_1 + a_2 x + a_3 y + a_4 xy \\
I_{\Phi_y}(x,y) &:= a_5 + a_6 x + a_7 y + a_8 xy \, .
\end{aligned}
\tag{4.50}
$$

To get the coefficients $\boldsymbol{a} = (a_1, ..., a_8)^T$ we solve the linear system of equations

$$
\boldsymbol{X}\boldsymbol{a} = \boldsymbol{f} \, .
\tag{4.51}
$$

For a patch around (x_i, y_i) as shown in Figure 4.8 $\boldsymbol{X}$ and $\boldsymbol{f}$ are given by

$$
\boldsymbol{X} =
\begin{bmatrix}
1 & x_1 & y_1 & x_1 y_1 & 0 & 0 & 0 & 0 \\
0 & 0 & 0 & 0 & 1 & x_1 & y_1 & x_1 y_1 \\
1 & x_2 & y_2 & x_2 y_2 & 0 & 0 & 0 & 0 \\
0 & 0 & 0 & 0 & 1 & x_2 & y_2 & x_2 y_2 \\
1 & x_3 & y_3 & x_3 y_3 & 0 & 0 & 0 & 0 \\
0 & 0 & 0 & 0 & 1 & x_3 & y_3 & x_3 y_3 \\
1 & x_4 & y_4 & x_4 y_4 & 0 & 0 & 0 & 0 \\
0 & 0 & 0 & 0 & 1 & x_4 & y_4 & x_4 y_4 \\
0 & 1 & 0 & y_i & 0 & 0 & 1 & x_i
\end{bmatrix}
\quad \text{and} \quad
\boldsymbol{f} =
\begin{pmatrix}
\Phi_x(x_1, y_1) \\
\Phi_y(x_1, y_1) \\
\Phi_x(x_2, y_2) \\
\Phi_y(x_2, y_2) \\
\Phi_x(x_3, y_3) \\
\Phi_y(x_3, y_3) \\
\Phi_x(x_4, y_4) \\
\Phi_y(x_4, y_4) \\
-k^2 \Phi(x_i, y_i)
\end{pmatrix} \, .
$$

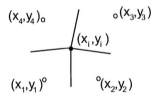

Figure 4.8: For the gradient interpolation at an interior node all adjacent centre gradients and the node related value for $\Delta\Phi(x_i, y_i)$ add information.

All gradients of the adjacent quadrangles (see Figure 4.8) and $\Delta\Phi(x_i, y_i)$ given by the solution of the Helmholtz equation are used to approximate the gradient at the interior node (x_i, y_i)

$$
\begin{pmatrix}
\Phi_x(x_i, y_i) \\
\Phi_y(x_i, y_i)
\end{pmatrix}
\approx
\begin{pmatrix}
I_{\Phi_x}(x_i, y_i) \\
I_{\Phi_y}(x_i, y_i)
\end{pmatrix}
=
\begin{pmatrix}
a_1 + a_2 x_i + a_3 y_3 + a_4 x_i y_i \\
a_5 + a_6 x_i + a_7 y_3 + a_8 x_i y_i
\end{pmatrix} \, .
\tag{4.52}
$$

Step 4: Linear interpolation for boundary node gradients

For almost every boundary node $(x_b, y_b) \in \delta V$ there are two adjacent quadrangle elements as shown in Figure 4.9.

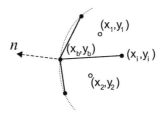

Figure 4.9: A boundary node (x_n, y_n) belonging to two quadrangle elements uses the averaged outer normal vector n, the gradients of the two adjacent centre points (x_1, y_1) and (x_2, y_2), and the gradient of the adjacent interior node (x_i, y_i).

For interpolating the gradients at such boundary nodes we define the linear interpolant

$$I_{\Phi_x}(x, y) := a_1 + a_2 x + a_3 y$$
$$I_{\Phi_y}(x, y) := a_4 + a_5 x + a_6 y. \tag{4.53}$$

To yield the coefficients we apply the gradients of the two adjacent centre points (x_1, y_1) and (x_2, y_2) and the gradient of the adjacent interior node (x_i, y_i). To compute the solution $a = (a_1, ..., a_6)^T$ we solve the linear system of equations

$$X a = f, \tag{4.54}$$

with

$$X = \begin{bmatrix} 1 & x_1 & y_1 & 0 & 0 & 0 \\ 0 & 0 & 0 & 1 & x_1 & y_1 \\ 1 & x_2 & y_2 & 0 & 0 & 0 \\ 0 & 0 & 0 & 1 & x_2 & y_2 \\ 1 & x_i & y_i & 0 & 0 & 0 \\ 0 & 0 & 0 & 1 & x_i & y_i \\ n_x & n_x x_b & n_x y_b & n_y & n_y x_b & n_y y_b \end{bmatrix}, \quad f = \begin{pmatrix} \Phi_x(x_1, y_1) \\ \Phi_y(x_1, y_1) \\ \Phi_x(x_2, y_2) \\ \Phi_y(x_2, y_2) \\ \Phi_x(x_i, y_i) \\ \Phi_y(x_i, y_i) \\ 0 \end{pmatrix}.$$

Within the entries of X we include the three adjacent gradients and the homogeneous Neumann boundary condition $\frac{\partial \Phi(x_b, y_b)}{\partial n} = 0$ at the boundary node. If the outer normal vector n has the zero component $n_x = 0$, then we compute the interpolant $I_{\Phi_x}(x, y)$ only and define $\Phi_y(x_b, y_b) = 0$, and vice versa, if the normal vector has the zero component $n_y = 0$.

If the boundary node (x_n, y_n) belongs to only one quadrangle element, then both derivative values are defined by zero values. If we compute the gradient of the velocity potential coupling modes $\nabla \Phi^C$ using the patch recovery technique, we have to consider the inhomogeneous Neumann boundary condition on $\Gamma_N \subset \delta V$.

Step 5: Approximation of the derivatives in the global SEM basis

Finally, we use the approximated gradients $\Phi_x(x, y)$ and $\Phi_y(x, y)$ as function values at the Q^2 quadrature points in every element to compute the global coefficients of the original mesh

$$M \hat{\Phi}_x = f_1, \qquad M \hat{\Phi}_y = f_2. \tag{4.55}$$

Here, M is the global mass matrix and f_1 and f_2 contain the information of the interpolated derivatives in x- and y-direction, respectively.

We repeat this patch recovery procedure for all velocity potential normal and coupling modes Φ^N and Φ^C. Consequently, we obtain the velocity normal modes v^N and the velocity coupling modes v^C using the velocity potential gradients as described in Definition 3.4. To get the pressure coupling modes p^C, we have to apply the patch recovery technique to the gradients of the velocity potential coupling modes $\nabla \Phi^C$. Now we have all normal and coupling modes in the same SEM formulation. This is advantageous for the coupling method (Section 3.1.2), because all computational work can be achieved with the help of the SEM. In three dimensions the application of the described patch recovery technique is straight forward.

4.4.2 Convergence

In the second part of this section we analyse the quality of the patch recovery technique. Exemplarily, we take the two dimensional rectangular geometry $V = [0, 1]^2$, where we know the analytical solution for the eigenfrequencies and eigenmodes for the pressure normal modes and for the velocity normal modes p^N and v^N (see Section 2.5.1). The coarse mesh on V consists of four skewed quadrangle elements, see Figure 4.10. In the following, we carry out the h-refinement and the p-refinement and compute the numerical error for the velocities v^δ in the energy norm

$$\epsilon_E := \left\| v - v^\delta \right\|_E = \left(\int_V |v(x) - v^\delta(x)|^2 \, dV \right)^{\frac{1}{2}}. \tag{4.56}$$

We will show the convergence for the velocity normal modes out of the velocity potential normal modes related to ω_{11}, ω_{22}, and ω_{13} using the patch recovery technique. For the other eigenfrequencies we yield similar results. The related reference velocity normal modes are computed by the same SEM formulation to obtain ϵ_E.

At first, we fix the number of quadrangle elements, as shown in the coarse mesh (Figure 4.10), and increase the polynomial degree P for the SEM basis expansion. We compute the velocity normal modes for $P = \{4, 5, 6, 7, 8, 9\}$ in every coordinate direction, which is related to the quadrature points $Q = \{6, 7, 8, 9, 10, 11\}$ in every coordinate direction. Figure 4.11 shows the refined mesh used by the patch recovery technique for $P = 7$ and the related number of quadrature points $Q = 9$ in each coordinate direction.

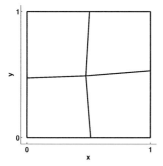

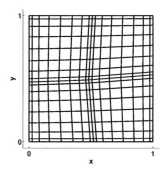

Figure 4.10: Coarse mesh with four skew elements on $V = [0,1]^2$.

Figure 4.11: Refined mesh with polynomial degree $P = 7$ $(Q = 9)$.

The numerical error in the energy norm ϵ_E for the p-refinement for each velocity normal mode related to the total number of degrees of freedom (N_{dof}) is plotted in Figure 4.12. Exemplarily for the velocity mode related to ω_{22} we observe that the general convergence rate is 0.86. We use $N_{dof} \sim P^2$ and conclude

$$\epsilon_E \sim P^{-2}. \tag{4.57}$$

For the h-refinement we fix the polynomial degree $P = 4$ $(Q = 6)$ and increase the number of quadrangle elements $N_{el} = \{4, 16, 64, 256\}$ by dividing each quadrangle into four new ones for every refinement step. Figure 4.13 presents the resulting error in the energy norm for the three considered velocity modes.

With the error in the energy norm for the velocity mode related to ω_{11} we observe that the general convergence rate is 0.53. We apply $N_{dof} \sim h^{-2}$, where h is the element length in the considered mesh, and conclude

$$\epsilon_E \sim h. \tag{4.58}$$

Using equation (4.57) and (4.58) we get

$$\epsilon_E = C\,h\,P^{-2}. \tag{4.59}$$

Here, C depends on the mesh, the angles in the quadrangle elements, and the variation of the velocity mode (see equation (4.27)). A similar convergence analysis can be carried out in three dimensions. The proof of the convergence result and the inclusion into the error estimates of the SEM is behind the scope of this work.

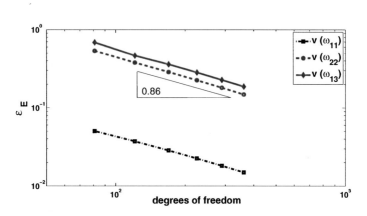

Figure 4.12: The error in the energy norm for the p-refinement with the patch recovery technique for $P = \{4, 5, 6, 7, 8, 9\}$ resulting in $\{81, 121, 169, 225, 289, 361\}$ degrees of freedom.

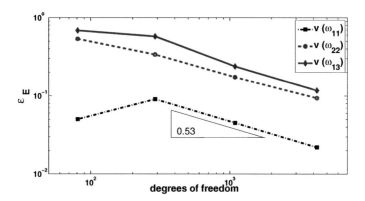

Figure 4.13: The error in the energy norm for the h-refinement with the patch recovery technique for $N_{el} = \{4, 16, 64, 256\}$ resulting in $\{81, 289, 1089, 4225\}$ degrees of freedom.

5 Numerical Simulations

In the following we apply the coupling method presented in Chapter 3 to various bench-mark examples. The frequency response, typically measured in dB, describes the magnitude of the system's response to the sound source. Thus the frequency response function (frf) is computed to analyse the sound field in the related system configuration. Consequently, we are able to investigate the influence of the systems geometry, the absorptive boundaries and the location and geometry of sound sources amongst others. All subsequent simulations are computed with a self-written code in MATLAB [76] called Room Acoustical Problem Solver (RAPS). RAPS will be used in [13, 101] and will be further developed at the Chair of Structural Mechanics at the Technische Universität München [84].

5.1 Point Impedance and Single Layer Absorber in 1d and in quasi-1d

5.1.1 Point Impedance in 1d

We begin with an example in the 1d domain $V = [0, L_x]$ including a sound source at $x = L_x$. The sound source excites the air with constant pressure and a circular frequency of excitation ω $(f_{Load}(x, t) = p_{Load}e^{i\omega t})$. At $x = 0$, an absorber with a local reacting point impedance (see Appendix (B.1)) is given by

$$Z(\omega) = i(\omega M - \frac{K}{\omega}) + C, \qquad M = 5\,kg, \;\; K = 2 \cdot 10^6 \frac{N}{m}, \;\; C = 600 \frac{Ns}{m}. \tag{5.1}$$

The reference solution in the frequency domain $p_{ref}(x, \omega) = \tilde{A}e^{-ikx} + \tilde{B}e^{ikx}$ is computed with two equations resulting out of the configuration constraints

$$(Z'(\omega) - 1)\,\tilde{A} \;\; - \;\; (Z'(\omega) + 1)\,\tilde{B} \;\; = \;\; 0, \qquad \tilde{A}, \tilde{B} \in \mathbb{C},$$

$$e^{-ikL_x}\,\tilde{A} \;\; + \;\; e^{ikL_x}\,\tilde{B} \;\; = \;\; p_{Load}.$$

Here, Z' denotes the scaled impedance

$$Z'(\omega) := \frac{Z(\omega)}{\rho_0 c_0}.$$

For the simulation we choose $L_x = 3\,m$, $p_{Load} = 1\,\frac{N}{m^2}$ and $\Omega \in [1\,Hz, 300\,Hz]$ sampled with $\Delta\Omega = 0.5\,Hz$ and considering $\Omega = \frac{\omega}{2\pi}$. The domain V is subdivided into eight equal elements each with $Q = 7$ supporting nodes and the basis functions have a polynomial degree of $P = 5$. Consequently every mode has 41 dofs. For the impedance boundary and the sound source at the boundary one coupling mode each is applied for the pressure and for the velocity, respectively. Further on, we include the first four fixed-interface normal modes without the zero frequency mode. First, we compute the coefficients A_r and B_s for the normal and coupling modes for every frequency of excitation. Then the solution is scaled due to the acoustic sound pressure level (SPL) corresponding to the amplified pressure of the sound source

$$\text{SPL}_{Amp}(\omega) := 10\log\left(\sqrt{\frac{1}{\int\limits_V dx}\int\limits_V \left|\frac{p(x,\omega)}{p_{Load}}\right|^2 dx}\right)\ [dB]. \tag{5.2}$$

In Figure 5.1 the frfs of the analytical solution (green) and the numerical solution (blue) are plotted corresponding to the frequency Ω. With the dynamic behaviour of the frf the influence of the point impedance (5.1) is observable. At the first resonance around $30\,Hz$ the stiffness influence is dominant. In the range of the second resonance ($90\,Hz$) we recognize the damping, caused by the damping coefficient C. Afterwards the amplitudes of the following resonances are growing again.

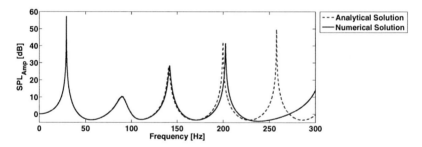

Figure 5.1: Comparison of the analytical and the numerical solution for the point impedance Z given in (5.1) using four normal modes and two coupling modes.

This damping influence of the complex impedance function $Z(\omega)$ is related to its eigenfrequency $(max\,|i\omega Z(\omega)|^{-2}$, see Appendix B.1), which is $99\,Hz$ for the chosen three impedance parameters in equation (5.1). The growing of the resonance amplitudes afterwards is typical and is caused by the mass parameter M. Consequently, it becomes clear that a local reacting point impedance, related to its own resonance, is tuned for a certain range of frequencies to pull out energy of the system. That means the amplitudes of the frf are reduced significantly in this specific frequency range.

Up to a frequency of $170\,Hz$ the numerical solution fits very well to the analytical one.

Afterwards the growing difference between both solutions is dependent on the number of applied normal modes. For the considered geometry here the related eigenfrequency of the forth normal mode is $229.85\,Hz$. To avoid such a significant numerical error we recommend to consider normal modes up to a eigenfrequency $\omega \leq 1.5\,\omega_{Load}$, where ω_{Load} is the frequency of excitation in each calculation, respectively.

5.1.2 Single Layer Absorber in quasi-1d

We consider the two dimensional domain $V = [0, L_x] \times [0, L_y]$. At $x = 0$, a single layer porous absorber with a thickness of $d = 0.1\,m$ made of polyurethane foam is placed at the wall along y. The related impedance $Z(\omega)$ is computed for this porous absorber as described in Section 3.3. At $x = L_x$ a sound source is applied exciting plane waves along y. Exemplarily we sketch this configuration with a microphone as the sound source in Figure 5.2.

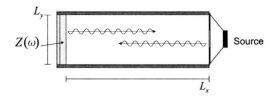

Figure 5.2: Set up of the quasi 1d - 2d configuration.

For the simulation we choose $L_x = 3\,m$, $L_y = 1\,m$, $p_{Load} = 1\,\frac{N}{m^2}$ and $\Omega \in [1\,Hz, 300\,Hz]$. Within the coupling method V is subdivided into eight equal elements with $Q = 7$ supporting nodes and the basis functions have a polynomial degree of $P = 5$. Because of the plane wave excitation again one coupling mode each is applied for the source and for the single layer absorber, respectively. Further on, we include the first eight fixed-interface normal modes without the zero frequency mode, for the pressure and the velocity, which is enough to gain a negligible numerical error. To analyse the influence of the single layer absorber qualitatively a second frf is computed with a totally reflective wall at $x = 0$ along y instead. Both frf solutions are plotted in Figure 5.3.

The result of this fluid-structure interaction example shows a shift of the natural eigenfrequencies and a reduction of the amplitudes of the natural eigenfrequencies ($28.7\,Hz$, $86.3\,Hz$, $143.7\,Hz$, $201.2\,Hz$, $258.7\,Hz$). The reduction of the amplitude is increasing with higher frequencies with respect to the amplitude of the first eigenfrequency $86.3Hz$. This is due to the working behaviour of a real absorber.

A real absorber is working more efficiently for higher frequencies, where the resulting wavelengths are shorter and therefore the sound velocity within the absorber compared to the maximum velocity is higher than for low frequencies [33, 37, 79].

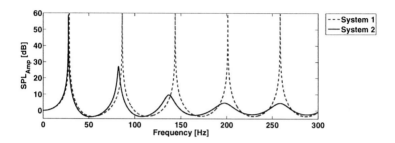

Figure 5.3: Comparison between the frfs of the acoustic volume V with the totally reflecting wall (System 1) and with the single layer absorber (System 2).

Remark:

For the limit case $Z = 0$ the pressure is forced to be zero at the left end, which equals to an open end. In this case the coupling mode for the impedance has to permit a velocity at the left and has to realize a pressure equal to zero. Meanwhile the normal modes for the velocity as well as the coupling mode for the velocity of the sound source at the right are zero at the left end. With a growing number of normal modes we achieve that the pressure at the left end converges to zero.

5.2 Coupling of Air Components

Considering the substructure of a complex system into more simple subsystems as presented in the acoustic network approach in Figure 3.1, it is necessary to couple air components with the coupling method. The following simulation examples neglect the inclusion of realistic boundaries and sound sources to focus on the air coupling only. Therefore we prescribe totally reflecting boundaries for the enclosed system within the coupling procedure.

Observing the acoustic network in Figure 3.1 it is common that the total air volume is divided into air components. These are connected by transmitting boundaries. To accomplish the coupling between two air components we compute the same number of coupling modes for the related interface for the first and for the second air component. Thereby for each coupling mode of the first air component the velocity amplitude at the interface has to be the same as the velocity amplitude of the related coupling mode in the second air component. We arrange the coupling then by only one set of solution coefficients for this specific interface. For that we rearrange the matrix M (see equation (3.12)) with the appropriate overlapping of the air components of the related submatrices. With this overlapping procedure every air component is connected directly or indirectly with all other air components by the matrix M.

5.2.1 Coupling of Two Air Components in 1d

The one dimensional configuration is built up by an air component 1 with a length of L_1 and a reflective boundary at the left end. This first component is coupled to a second air component with a length L_2 and a reflective boundary at the right end. The Coupling of both components is realized with a coupling mode each to enforce the velocity at the interface to be equal. The pressure continuity is forced indirectly within the energy equilibrium with the help of Hamilton's Principle as defined in equation (3.11). Because there is no sound source we gain a general eigenvalue problem in ω^2. The components inside the resulting eigenvectors are the modal parameters contributing to the related eigenfrequency.

In the following simulation we define $L_1 = 8\,m$ and $L_2 = 5\,m$. In addition we choose eight normal modes in the first component ($n_1^N = 8$) and five normal modes in the second component ($n_2^N = 5$). In Figure 5.4 we present the first, the second and the forth normalized pressure eigenmode shapes. Obviously there is a pressure discontinuity at the interface between both components ($x = 0$), which is growing with an increasing related eigenfrequency.

This is due to the fact that the pressure continuity cannot be enforced with the sum of normal and coupling modes (see Section 3.1). In addition the occuring jump at the interface increases with a related higher eigenfrequency ω. To see the convergence behaviour for increasing the number of applied normal modes we analyse two cases. In the first case equal sized components are used ($L_1 = L_2$) and the number of normal modes in component 1 is equal to the number of normal modes in component 2 ($n_1^N = n_2^N$). In the second

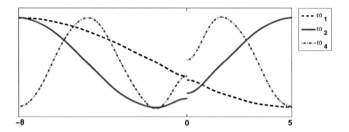

Figure 5.4: First, second and fourth pressure eigenmodes of the coupled system with a discontinuity at the interface ($x = 0$).

case non-equal sized components are applied. To guarantee the best approximation for the solved eigenfrequencies and eigenmodes of the coupled system in the second case the number of normal modes in component 1 and the number of normal modes in component 2 underlie the relation

$$n_1^N = n_2^N \cdot round\left(\frac{L_1}{L_2}\right) . \tag{5.3}$$

In Figure 5.5 and 5.6 the relative error of the first ten eigenfrequencies ω_i to the corresponding eigenfrequencies of the total system with a equal discretization are analysed for both cases with an increasing number of normal modes each. We observe that for a fixed eigenfrequency ω_i and an increasing number of normal modes the relative error in the eigenfrequency is decreasing in both cases. In addition we recognize an oscillating behaviour of the relative error for a fixed number of normal modes in Figure 5.5.

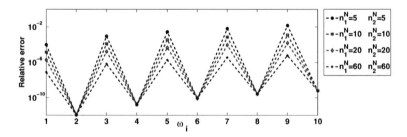

Figure 5.5: Relative error in the eigenfrequencies ω_i for equal sized air components ($L_1 = L_2 = 5\,m$).

This is because every second eigenmode of the coupled system is built up by exactly one eigenmode in each component. For the second case in Figure 5.6 there is no oscillat-

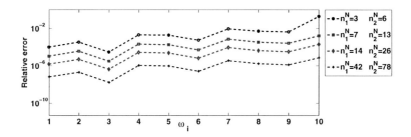

Figure 5.6: Relative error in the eigenfrequencies ω_i for non-equal sized air components ($L_1 = 3.5\,m$, $L_2 = 6.5\,m$).

ing but a nearly monotonic increasing relative error of the eigenfrequency, because every eigenmode of the coupled system needs more eigenmodes in each component.
In Figure 5.7 and 5.8 the pressure discontinuity at the coupling interface relative to the maximal absolute amplitude of the related eigenmode for the first ten eigenfrequencies of the coupled system are analysed for case 1 and 2 with an increasing number of normal modes each.

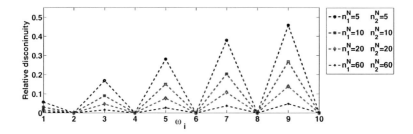

Figure 5.7: Relative pressure discontinuity at the interface for equal sized air components ($L_1 = L_2 = 5\,m$).

Figure 5.7 shows a regular oscillating behaviour of the pressure discontinuity for equal sized components (case 1) for a fixed number of normal modes, which is due to the previous mentioned reason. For case 2 in Figure 5.8 we gain a more regular increasing relative pressure discontinuity for increasing eigenfrequencies ω_i at the interface. The relative pressure discontinuity for a fixed number of normal modes is partly smaller than for the related relative pressure discontinuity in case 1.
Summarizing the results of the previous simulation examples we recommend to use nearly equal sized components. This results in more regular increasing in relative errors of the

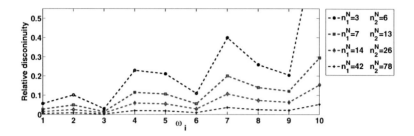

Figure 5.8: Relative pressure discontinuity at the interface for non-equal sized air components ($L_1 = 3.5\,m$, $L_2 = 6.5\,m$).

eigenfrequencies and in relative pressure discontinuities.

5.2.2 Coupling of Two Air Components in 2d

In the following we analyse the coupling in two dimensions. The first case treats two equal sized rectangular air components with $L_{x1} = L_{x2}$ and $L_{y1} = L_{y2}$ in x- and y-direction, respectively. The second case considers two non-equal sized rectangular components with the same length in y-direction ($L_{y1} = L_{y2}$). To guarantee the best approximation for the second case (see relation (5.3)) the number of normal modes in component 1 (n_1^N) and the number of normal modes in component 2 (n_2^N) underlie the following relation

$$n_1^N = n_2^N \cdot round \left(\frac{L_{x1} L_{y1}}{L_{x2} L_{y2}} \right) . \tag{5.4}$$

The coupling between both components along y needs only some coupling modes, because of the modal approach described in Section 3.2.3. For these rectangular components the number of coupling modes depends on the wavenumber in y-direction of the related eigenfrequency. Exemplary for a frequency ω_{ij} with the y-part of the wavenumber $\frac{j\pi}{L_{y1}}$ results that j coupling modes are necessary. For in general non-rectangular components the approximation becomes better with an increasing number of coupling modes.

For the simulations with the coupling method both components are discretized by squared elements with a length of $0.1\,m$ and the length in y-direction is $L_{y1} = L_{y2} = 1.0\,m$. The polynomial degree of the SEM basis functions is $P = 3$ and the related number of quadrature points is $Q = 5$. For the first case we choose $L_{x1} = L_{x2} = 1.1\,m$ and the resulting number of global dofs in both components is 1054. For the second case 2 we consider $L_{x1} = 1.1\,m$ and $L_{x2} = 1.3\,m$ and the resulting number of dofs in component two is 1240. The reference system for each configuration has 2077 and 2263 dofs, respectively.

In Figure 5.9 and 5.10 the relative error in ω_i compared to the related eigenfrequency in the reference system is presented for the corresponding case.

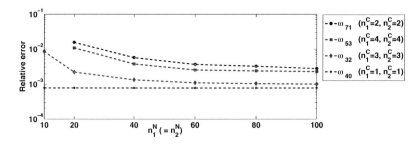

Figure 5.9: Relative error in the eigenfrequencies ω_{ij} for equal sized air components ($L_{x1} = L_{x2} = 1.1\,m$, $L_{y1} = L_{y2} = 1.0\,m$).

In Figure 5.9 (case 1) ω_{40} is already completely build up by the lowest number of normal modes in each component related to symmetry. The same results for every even wavenumber in x-direction (see also Figure 5.5 for the corresponding case in 1d).

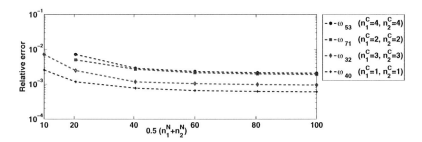

Figure 5.10: Relative error in the eigenfrequencies ω_{ij} for non-equal sized air components ($L_{x1} = 1.1\,m$, $L_{x2} = 1.3\,m$, $L_{y1} = L_{y2} = 1.0\,m$).

In Figure 5.10 (case 2) this eigenfrequency needs more normal modes to be built up, but is still the one with the lowest relative error. This is because ω_{40} is the smallest out of the four selected eigenfrequencies. The eigenfrequencies ω_{71} and ω_{53} are too high to be tackled with the first number of normal modes. Consequently we leave out these values in both simulation cases. Apart from the mentioned symmetry effects we observe that the convergence is very similar in both cases.

While analyzing the related pressure eigenmodes, there occur discontinuities along the interface between both components. This is the same phenomenon as for the one dimensional coupling of two air components considered in Section 5.2.1. The resulting pressure discontinuity is due to the continuous, but not continuous differentiable, coupling of velocity. The relative pressure discontinuities are computed with the ratio of the maximal absolute pressure discontinuity along the interface and the maximal absolute pressure amplitude of both components.

Subsequently, Figure 5.11 and 5.12 present the relative pressure discontinuities of the related eigenfrequency ω_{ij} along y for case 1 and 2.

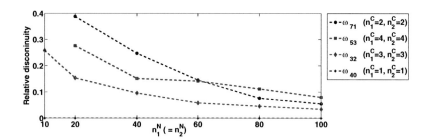

Figure 5.11: Relative pressure discontinuity at the interface for equal sized air components ($L_{x1} = L_{x2} = 1.1\,m$, $L_{y1} = L_{y2} = 1.0\,m$).

For the equal sized air components in Figure 5.11 (case 1) the relative pressure discontinuity seems to decrease faster with an increasing number of normal modes than in Figure 5.12 (case 2).

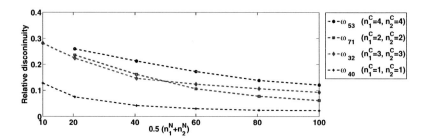

Figure 5.12: Relative pressure discontinuity at the interface for non-equal sized components ($L_{x1} = 1.1\,m$, $L_{x2} = 1.3\,m$, $L_{y1} = L_{y2} = 1.0\,m$).

Summarizing the previous analysed simulations of coupled two dimensional rectangular air components we suggest to use as similar sized components as possible in general to achieve a fast convergence in the relative error of the eigenfrequencies and in the pressure discontinuities, which is accomplished with an increasing number of normal modes.

5.3 Transmission Problem

If two air volumes are separated by a thin structure, which is able to vibrate, then sound waves can be transmitted from one volume to the other. In general an exciting sound source is positioned in one volume to produce a certain sound field. The resulting sound field forces the interconnecting structure to vibrate and to transmit energy to the second air volume. In the second air volume this structure-borne sound produces a specific sound field as well. This kind of transmission problems is a typical topic in standard literature [28, 32, 33, 37] and combines the coupling of air volumes as well as the inclusion of absorptive boundaries as presented in Section 5.1 and 5.2.

Plane waves through a partition in 1d

The configuration of the acoustic system consists of a component with $x_1 \in [-L_1, 0]$, where the sound source is placed at the left ($x_1 = -L_1$). The connecting partition is located at $x_1 = 0$. This partition is assumed to be infinite thin and has a specific impedance $Z(\omega)$. Therein the thickness of the partition is considered with the mass per surface area in the so-called surface density. Consequently this first air component with the sound source and the impedance boundary is similar to the system in Section 5.1. The second air component with $x_2 \in [0, L_2]$ is connected to the partition at $x_2 = 0$ and possesses a totally reflecting end at $x_2 = L_2$. For the applied coupling method (described in Chapter 3) we compute two coupling modes for the first component, one for the source at the left end and one for the impedance at the right end. In addition we compute one coupling mode for the second air component, which provides the coupling to the interconnecting structure. Because of the thin vibrating partition we require the coupling condition for the velocity at the interface

$$v_1(x_1)|_{x_1=0} = v_2(x_2)|_{x_2=0} . \tag{5.5}$$

Here, v_1 and v_2 are the velocities in component 1 and component 2. Consequently, the coefficients of the second coupling mode in component 1 and in front of the coupling mode in component 2 are enforced to be equal. This procedure is equal to the coupling of two air components as described in Section 5.2. Additionally we include the impedance $Z(\omega)$ of the interconnecting thin structure by the Lagrangian of the boundary condition as presented in equation (3.11). For this 1d benchmark problem an analytical reference solution can be calculated with the help of a condition for the pressure difference caused by the partition. This condition due to Newtons second law [33] is defined by

$$p_1(x_1)|_{x_1=0} - p_2(x_2)|_{x_2=0} = \frac{v_2(x_2)|_{x_2=0}}{i\omega} Z(\omega) . \tag{5.6}$$

We apply the point impedance $Z(\omega)$ (equation (5.1)), which is representative for a thin plate. The partition related parameters M, K and C are mass, stiffness and damping per unit area, respectively. More details about acoustic and mechanical impedance boundaries are given in Appendix B. The standard parameter to analyse transmission problems is the transmission ratio τ expressing the relation of the transmitted to the incident intensity. The transmission ratio is defined by

$$\tau = \frac{4\frac{\rho_1 c_1}{\rho_2 c_2}}{\left(\left(\frac{\omega M - \frac{K}{\omega}}{\rho_2 c_2}\right)^2 + \left(\frac{\tilde{\omega} M \eta}{\rho_2 c_2} + \frac{\rho_1 c_1}{\rho_2 c_2} + 1\right)^2\right)} \, . \tag{5.7}$$

The subscripts 1 and 2 denote the relation to the air components 1 and 2. In addition $\tilde{\omega} = \sqrt{\frac{K}{M}}$ represents the natural circular frequency of the partition structure and η is the in vacuo loss factor ($\eta = \frac{C}{\tilde{\omega} M}$). The related logarithmic index of sound transmission is the sound reduction index $R = 10 log(\frac{1}{\tau}) \, [dB]$ [33].
For the simulation examples we assume air with equal material parameters in both components ($\rho_1 = \rho_2 = \rho_0$, $c_1 = c_2 = c_0$). The transmission ratio τ is maximal for the natural circular frequency of excitation $\omega = \tilde{\omega}$. That means if the sound source is exciting the air in component 1 with the natural frequency of the partition structure almost all sound energy is transmitted, if $\eta << \frac{\rho_0 c_0}{\tilde{\omega} M}$. This phenomenon is known as the coincidence effect. If on the one hand $\omega << \tilde{\omega}$, then the transmission is mainly dependent on the stiffness $\tau \approx \frac{2\rho_0 c_0 \omega}{K}$. If on the other hand $\omega >> \tilde{\omega}$, then the transmission is mainly dependent on the mass $\tau \approx \frac{2\rho_0 c}{\omega M}$. We consider the partition to be a wooden-like panel with the impedance

$$Z(\omega) = i(\omega M - \frac{K}{\omega}) + C, \qquad K = 10^9 \frac{N}{m^2}, \, M = 160 \frac{kg}{m^2}, \, C = 0 \, . \tag{5.8}$$

The related natural eigenfrequency is $397.89 \, Hz$. In Figure 5.13 we present the transmission ratio τ and the sound reduction index R.

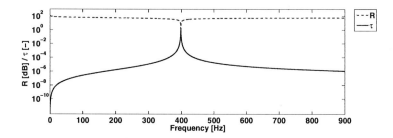

Figure 5.13: Pressure in component 1 (p_1) and pressure in component 2 (p_2) for a frequency of excitation $\Omega = 394.70 \, Hz$.

At the natural frequency of the partition we observe the typical values of the coincidence effect ($\tau = 1$, $R = 0$). We define the dimensions of the two components as $L_1 = L_2 = 5 \, m$ and the sound source excites the air of the first component with $p_{Load} = 1 \frac{N}{m^2}$.

In the following figures, frequencies of excitations around the natural frequency are applied to simulate the transmission problem of the considered configuration. Therein we use a fixed sufficient number of normal modes for each component to neglect the numerical error in relation to the analytical solution. In Figure 5.14 we show the pressure for the state $\omega < \tilde{\omega}$. We observe that a part of the sound energy is transmitted by the partition. Therefore a pressure wave possessing a smaller amplitude as in component 1 is induced inside the second air component.

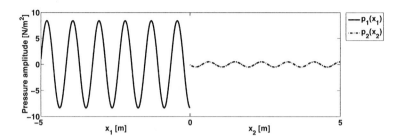

Figure 5.14: Pressure in component 1 (p_1) and pressure in component 2 (p_2) for a frequency of excitation $\Omega = 394.70\,Hz$.

In Figure 5.15 we present the coincidence effect for $\omega = \tilde{\omega}$. Obviously, all sound energy is transmitted from the left to the right air component and there is no pressure discontinuity at the partition anymore.

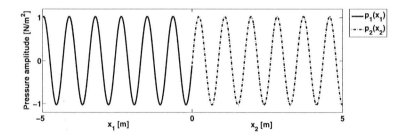

Figure 5.15: Pressure in component 1 (p_1) and pressure in component 2 (p_2) for a frequency of excitation $\Omega = 397.89\,Hz$ equal to the natural frequency of the partition (coincidence effect).

In Figure 5.16 and 5.17 we present the pressure solutions caused by circular frequencies of excitation with $\omega > \tilde{\omega}$ each. We observe that the amplitude of the pressure wave in component 2 is decreasing. This is because the partition is transmitting less energy for higher frequencies of excitation $(\tau \approx \frac{2\rho_0 c_0}{\omega M})$.

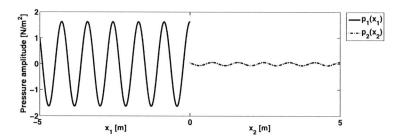

Figure 5.16: Pressure in component 1 (p_1) and pressure in component 2 (p_2) for a frequency of excitation $\Omega = 404.25\,Hz$.

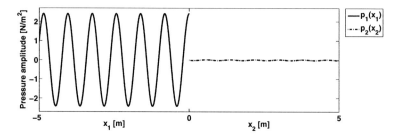

Figure 5.17: Pressure in component 1 (p_1) and pressure in component 2 (p_2) for a frequency of excitation $\Omega = 426.54\,Hz$.

Remark:

The amplitudes of the pressure solution in component 2 are in general neither monotonically increasing before nor monotonically decreasing after the natural circular frequency $\tilde{\omega}$ of the partition. This is because component 2 is a closed cavity and therefore natural frequencies of this enclosure can be excited with the vibrating partition as well. In two and three dimensions the transmission problem can be applied similar within the coupling method and additionally, we have to consider the angles of incident corresponding to the sound waves.

5.4 Car Interior FSI in 2d

In this section we present an FSI problem in a two dimensional car interior (Figure 5.18). The geometry for the car is taken out of [97]. After a discretization into 123 quadrangle elements we compute the velocity normal modes for a polynomial degree $P = 3$ and $Q = 5$ quadrature points in each coordinate direction resulting in 1207 dofs of the total system.

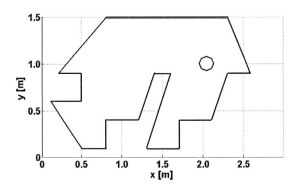

Figure 5.18: Geometry of the car interior with a single layer absorber at the roof and a circular sound source above the back seat.

The corresponding eigenfrequencies fit to the stated eigenfrequencies in [97]. In Figure 5.19 to 5.22 we present the second, the third, the 10th and the 30th normalized velocity potential normal mode, respectively. Obviously, the dynamic behaviour of the eigenmode shape is increasing for higher eigenfrequencies.

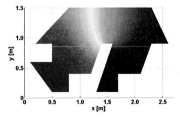 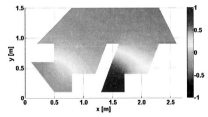

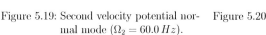

Figure 5.19: Second velocity potential nor-   Figure 5.20: Third velocity potential normal mode ($\Omega_2 = 60.0\,Hz$).   mal mode ($\Omega_2 = 114.4\,Hz$).

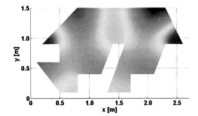
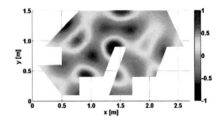

Figure 5.21: $10th$ velocity potential normal mode ($\Omega_2 = 192.5\,Hz$).

Figure 5.22: $30th$ velocity potential normal mode ($\Omega_2 = 583.6\,Hz$).

We obtain the velocity normal modes by applying the Patch Recovery Technique, described in Section 4.4. Then, we compute the pressure normal modes by using the velocity potential normal modes and the related eigenfrequencies as presented in equation (3.20). Within this enclosure we want to analyse the effect of a plate-like absorber at the roof of the car (see Figure 5.18). Therefore we place a single layer absorber at $y = 1.5\,m$ along y made by the material polyurethane foam (see Section 5.1.2). To compute the velocity potential coupling modes we enforce the beam modes, described in Section 3.3, as inhomogeneous boundary conditions along the predefined absorptive boundary. In Figure 5.23 to 5.26 we present the first, the second, the third, and the fourth velocity potential coupling mode.

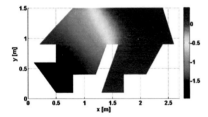
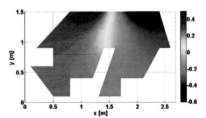

Figure 5.23: First coupling mode.

Figure 5.24: Second coupling mode.

To obtain the velocity and the pressure coupling modes, we apply the Patch Recovery Technique again.

We consider a harmonically oscillating circular sound source located at $x = 2\,m$ and $y = 1\,m$ with a radius of $0.1\,m$ (see Figure 5.18). This position is similar to the mouth of a person sitting at the back seat of the car. This sound source excites the interior sound field with a constant pressure $p_{Load} = 1\,\frac{N}{m^2}$.

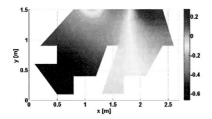

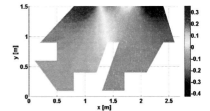

Figure 5.25: Third coupling mode..   Figure 5.26: Fourth coupling mode.

We compute the frf up to the frequency of excitation $210\,Hz$ with $\Delta\Omega = 1\,Hz$ applying 15 normal and three coupling modes. Comparing the resulting scaled sound pressure level (equation (5.2)) in Figure 5.27 with the first eigenfrequencies of the totally reflecting car (Figure 5.19 and 5.20), we notice that the resonances are shifted to the left and damped in amplitude due to the single layer absorber at the roof.

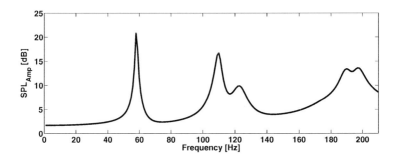

Figure 5.27: frf of the harmonically excited car interior sound field up to the frequency of excitation $\Omega = 210\,Hz$.

93

6 Uncertainty Analysis

Different physical realizations of a manufactured product have physical and geometric properties which inevitably differ in detail. A number of factors, including normal manufacturing variations, environmental conditions, and variations in operating conditions, cause this variability. The inherent variability in the properties of the system consequently produces variability in the dynamic response. This dynamic response can for example be calculated with the frequency response function (frf).

In general there are two types of uncertainty, the reducible and the irreducible uncertainty. Reducible uncertainty stands for lack of knowledge about input data and the behaviour of a system. Irreducible uncertainty is due to varying environmental and system related parameters. Next to this, a differentiation can be made into parametric and non-parametric uncertainties, the former directly related to given parameters and the latter often associated with all other effects, such as the accuracy of the model. The numerical model of a real system, including the defined parameters, is always an idealized system. It is therefore impossible to tackle all non-deterministic effects by considering parametric uncertainties only. However, in many cases it is reasonable to quantify variation in the input parameters only. In the last years, numerical methods considering uncertain parameters have become more and more of interest [44, 73].

In the following, we will analyse methods considering parametric uncertainties related to the room acoustical problems handled in Chapter 2. We choose these methods in dependency of our coupling method described in Chapter 3 for the low- and mid-frequency range. Therefore we neglect methods embedded in the Statistical Energy Analysis (SEA) [58, 60, 70] for the high-frequency range.

Therein, restricting ourselves to irreducible uncertainty, two kinds of methods handling these problems have to be distinguished: the possibilistic approach and the probabilistic approach, which are described for example in [49]. This irreducible uncertainty can be caused by changes in the geometry of the cavity, in the position or the geometry of the absorber or the load, as well as in environmental effects like the density or the temperature of the air. In Section 6.2.1 and 6.2.2 we focus on varying parameters of the absorber. Basic possibilistic and probabilistic concepts are applied on benchmark configurations to estimate the variation in frequency response functions from uncertain input data.

6.1 Characterization of Uncertainty and Methods

In this section we want to give a general insight into uncertainty analysis. Therefore, we define and describe the different kinds of uncertainty first. Then we introduce the approaches to represent these kinds of uncertainty and at the end of this section we give an overview of the methods to treat uncertainty.

6.1.1 Types of Uncertainty

Uncertainty is a term used in subtly different ways and in many various fields.

Definition 6.1. Uncertainty
Uncertainty describes the vagueness and indetermination of physical measurements already made, predictions of future events, or the unknown.

Further, some quantities vary over time and over space. This is termed variability. We interpret variability as uncertainty under certain conditions. However, the implications of the differences in uncertainty and variability are relevant in decision making. Environmental systems are inherently stochastic due to unavoidable randomness.

Definition 6.2. Natural uncertainty
Natural uncertainty occupies random quantities in principle. In addition it includes quantities, who are precisely measurable, but are modelled as random quantities as a practical matter due to measurement costs.

These natural uncertainties are material parameter like density or geometry. Non-parametric uncertainty can be divided in two types: model uncertainty and parametric and data uncertainty.

Definition 6.3. Model uncertainty
Mathematical models can be an approximation of the real system. The limited spatial or temporal resolution of many models is also a type of approximation.

Model uncertainty subdivides into uncertainty related to the model structure, the details of the model, the extrapolation of input data, the resolution, and the model boundaries.

Definition 6.4. Parametric and data uncertainty
Uncertainties in model parameter arise from estimations, measurement errors, and misclassifications.

Uncertainty associated with model formulation and application can also be classified as reducible and irreducible uncertainty.

Definition 6.5. Irreducible and reducible uncertainty
The irreducible uncertainty in data and models is generally a result of the presence of natural uncertainty. Reducible uncertainty can be decreased, for example, by developed methods, improved instrumentation, and improvements in model formulation.

6.1.2 Approaches for Representation of Uncertainty

Some of the widely used uncertainty representation approaches include interval mathematics, fuzzy theory, and probabilistic analysis [2]. These approaches are presented in the following.

Possibilistic approach (interval mathematics):
Possibilistic approaches [6, 80] are used to address data uncertainty that arises on the one side due to imprecise measurements and on the other side due to the existence of several alternative methods, techniques, or theories to estimate model parameters. In many cases, it may not be possible to obtain the probabilities of different values of imprecision in data. In some cases only error bounds can be obtained. This is especially true in case of conflicting theories for the estimation of model parameters, in the sense that probabilities cannot be assigned to the validity of one theory over another. In such cases, interval mathematics can be used for uncertainty estimation, as this method does not require information about the type of uncertainty in the parameters. The objective of interval analysis is to estimate the bounds on various model outputs based on the bounds of the model inputs and parameters. In the possibilistic approach, uncertain parameters are assumed to be unknown but bounded. Therefore each uncertain parameter has upper and lower limits. The goal of a possibilistic propagation approach [31] is to calculate the bounds on the response quantity of interest. The primary advantage of possibilistic methods is that they can address problems of uncertainty analysis that cannot be studied through probabilistic analysis. However, these methods do not provide adequate information about the nature of output uncertainty, as all the uncertainties are forced into one arithmetic interval. Especially when the probability structure of input data is known, the application of interval analysis would in fact ignore the available information, and hence is not recommended.

Fuzzy theory:
Fuzzy theory is a method that facilitates uncertainty analysis of systems where uncertainty arises due to vagueness or fuzziness rather than due to randomness alone [6, 59]. This is based on a superset of conventional logic that has been extended to handle the concept of partial truth. In the classical set theory, the truth value of a statement can be given by a membership function. This function allows either the value true (1) or the value false (0). In fuzzy theory, statements are described in terms of membership functions, that are continuous and have a range [0, 1]. Fuzzy arithmetic is based on the grade of membership of a parameter in a set to calculate the grade of membership of a model output in another set. Therefore, we consider Fuzzy theory as a generalization of the classical set theory. However, Fuzzy theory appears to be more suitable for qualitative reasoning, and classification of elements into a fuzzy set, than for quantitative estimation of uncertainty.

Probabilistic analysis:
In the probabilistic approach, uncertainties are characterized by the probabilities associated with events. The probability of an event can be interpreted in terms of the frequency of occurrence of that event. Probabilistic analysis is the most widely used method for characterizing uncertainty in physical systems, especially when estimates of the probability distributions of uncertain parameters are available. This approach describes uncertainty arising from stochastic disturbances, variability conditions, and risk considerations. Therein, the uncertainties associated with model inputs are described by probability distributions, and the objective is to estimate the output probability distributions. The uncertainty in the parameters is specified by a probability density function (pdf), with a mean value and a standard deviation. Similarly, the variation in the response can be quantified in terms of distribution functions or statistics.

In practice there is often not enough data to quantify a distribution exactly and a standard pdf, such as a normal distribution, is assumed. The mean value can be taken as the deterministic value and only the variance has to be quantified. A normal distribution is often a reasonable assumption to model variability in physical processes. In statistics, this is also supported by the Central Limit Theorem [35], which states that any sum of many independent and identically distributed variables with finite variance is approximately normally distributed. The unbounded tails of the normal distribution are often inconsistent with reality, which has to be taken into account. Subsequent decomposition schemes, such as the Karhunen-Loève expansion [44], lead to a system of random algebraic equations, which are accessible by uncertainty propagation methods.

In contrast to sampling approaches, there are various subspace projection schemes [48], such as polynomial chaos expansion [43] and stochastic reduced basis methods [49]. This process comprises of two stages. In the first stage the probabilistic distribution of the input parameters has to be determined. In the second stage the uncertainty has to be propagated through the model.

6.1.3 Sensitivity and Uncertainty Analysis Methods

In probabilistic approaches, information about the likelihood and probability of events are included. The variation in the uncertain parameter is specified by a probability density function (pdf), with mean value and standard deviation. Similarly, the variation in the response can be quantified in terms of distribution functions or statistics.

Sampling methods:
The Monte Carlo (MC) method [36, 43] is a widespread standard for propagating probabilistic data. In standard MC sampling, parameter values are randomly drawn according to their probability distributions and a deterministic problem is solved for each sample. The method is very robust and converges with $N^{-\frac{1}{2}}$ to the exact solution, N being the sample size. It makes no approximations and considers all effects modelled in the deterministic problem. However, the numerical cost to estimate a small probability of failure can be in the order of thousands of deterministic solutions.

Subspace projection schemes:
In contrast to sampling approaches, there are various subspace projection schemes [95], such as polynomial chaos expansion [38, 73] and stochastic reduced basis methods [82]. These methods are based on constructing the functional dependence of the solution s as given by

$$s(\xi) = \sum_{n=0}^{\infty} s_n \Psi_n(\xi). \tag{6.1}$$

Here, Ψ_n denotes a suitably functional of the random variables and s_n stands for the deterministic coefficients.

The notion of the Polynomial Chaos (PC) expansion was introduced as a generalization of Fourier series expansion. The objective was to investigate the utility of Hermite polynomials in the integration theory with respect to Brownian motion. In the frame of the generalized PC approach it is possible to employ basis functions from the Askey family of orthogonal polynomials. The Hermite chaos expansion appears as a special case in this generalized approach, which is also referred to Wiener-Askey chaos. The motivation for generalized PC expansions arises from the observation that the convergence of Hermite chaos expansions can be far from optimal for non-Gaussian inputs. In such cases, the convergence rate can be improved by replacing Hermite polynomials with other orthogonal polynomials that best represent the input. Convergence results of the PC expansion applied to second-order stochastic processes are used to derive subspace projection schemes for stochastic FE analysis later on.

Local approximation techniques:
Local approximation techniques such as perturbation, Taylor expansion, Neumann series, and sensitivity based methods offer computationally efficient alternatives to the MC techniques and have been successfully applied to compute the first two statistical moments of the response quantities [48]. However, the major drawback of these local approximation techniques is that the results become inaccurate when the coefficients of variation of the input random variables are increased. The perturbation method uses a Taylor series expansion of the quantities involved in the equilibrium equation of the system around their mean values. The limitation of the perturbation method is that it requires derivatives with respect to the random variables which are often cumbersome and expensive to compute. It is often intricate and time consuming to obtain approximations beyond first-order. Hence, the perturbation method is generally applicable only for very small coefficients of variation.

Response surface techniques:
The Response surface method (RSM) aims at fitting a function to the model data. In practice, the method of least squares is used to fit a polynomial model which is then used to predict the response for the original model. A non-intrusive PC projection scheme can be applied to compute the undetermined coefficients when the basis is spanned by multidimensional Hermite polynomials in uncorrelated Gaussian random variables.

6.2 Parametric Uncertainty in Room Acoustics

In the following we will consider parametric uncertainty for the room acoustic problems of interest (Chapter 2). Therefore we have to consider the uncertainty of the input data for the procedure of the coupling algorithm (Section 3.1.3). In addition, we choose an appropriate method to propagate the uncertainty through the coupling method. At the end we analyse the statistics in the output data, for example the different samples of frfs. Figure 6.1 shows the additional steps for the input and output data for considering uncertainty in our coupling method presented in Chapter 3.

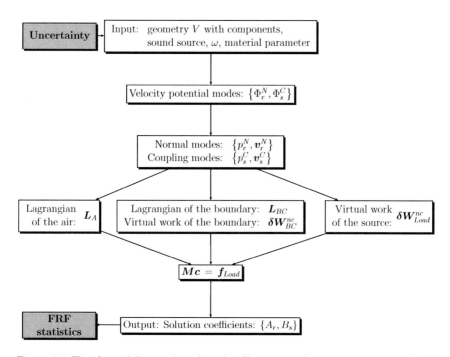

Figure 6.1: Flowchart of the coupling algorithm (Section 3.1.3) with additional steps (red) for uncertain input data and uncertainty of the output data.

To choose an appropriate method to propagate the uncertainty through the coupling method, we have to analyse the different possibilities of uncertain input data and their effects for the coupling procedure in detail. In Table 6.1 we list all possible uncertainties

of our room acoustical system we want to consider. On the right-hand side of the table
we remark the related affected part of the coupling methods linear system of equations
(see equation (3.12)).

Table 6.1: Sources of uncertainty in our room acoustical systems and the related affected
part of the coupling method.

Source of uncertainty	Affected part of the coupling method
Geometry of the acoustic volume V	Lagrangian of the air $\boldsymbol{L}_A$ (by the normal modes $\{p^N, \boldsymbol{v}^N\}$)
ρ_0, c_0 (material parameter of the air)	Lagrangian of the air $\boldsymbol{L}_A$ (by a linear factor in the pressure modes $\{p^N, p^C\}$ as given in Definition 3.3)
$Z(\omega)$ (impedance caused by parameters of the absorber like mass, stiffness, thickness (of the layers), and porosity)	Lagrangian of the boundary $\boldsymbol{L}_{BC}$ and virtual work of the boundary $\boldsymbol{\delta W}_{BC}^{nc}$
Absorber (geometry and location)	Lagrangian of the air (by the coupling modes $\{p^C, \boldsymbol{v}^C\}$
Sound source (type, geometry and location)	Virtual work of the forces $\boldsymbol{\delta W}_{Load}^{nc}$

The methods, introduced in Section 6.1.3, to propagate uncertainty have been analysed
in the frame of CMS methods, for example in [5, 49].
In the next sections, we want to apply some of these methods and analyse the statistics
of the output data by the corresponding frfs.

6.2.1 Possibilistic Approach

When there is no distribution given for uncertain input data, it is common to choose the possibilistic representation for the data. The set of possibilistic representations (see Section 6.1.2) includes, amongst others, interval analysis, convex modelling, and fuzzy set approaches [49, 80]. Because of nescience of the distribution of the uncertain variables, only lower and upper limits are possible. Therefore an interval for each variable can be given as input data.

We apply the interval analysis to the set up described in Figure 6.2, where the impedance at $x = 0$ is defined as a point impedance $Z(\Omega)$ in dependency of mass, stiffness and damping, as in equation (6.2). See Appendix B for more information.

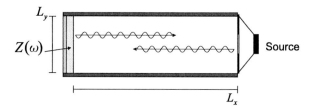

Figure 6.2: Set up of the 2d geometry configuration.

The mass M, the stiffness K and the damping coefficient C are the uncertain variables, which, for academic reasons, are chosen in huge intervals in this simulation example.

$$Z(\omega) = i\left(\omega M - \frac{K}{\omega}\right) + C, \qquad M \in [5\,kg,\,15\,kg],$$

$$K \in \left[10^5\,\tfrac{N}{m},\,4\cdot10^5\,\tfrac{N}{m}\right], \qquad C \in \left[600\,\tfrac{Ns}{m},\,800\,\tfrac{Ns}{m}\right] \tag{6.2}$$

In the frame of interval methods, the first choice is to compute all possibilities of interval endpoint combinations (2^3 combinations in this example) for each frequency of excitation. The resulting possible frfs can be described by a lower and upper frf envelope. In Figure 6.3 the resulting upper and lower frf envelopes for the acoustic sound pressure level SPL_{Amp} (equation 5.2) are presented. In a second step three additional equal spaced values inside each interval are considered. Figure 6.4 shows the more refined results now regarding 5^3 simulations per frequency.

The biggest differences between the envelopes in Figure 6.3 and 6.4 are visible at the resonances around $30\,Hz$, $85\,Hz$ and $145\,Hz$. Instead of equal spaced values, also other methods are possible to refine the envelopes. In general these methods are only applicable, if the intervals of uncertainty are very small and the system is not that sensitive to changes in the related parameters. In case of strongly non-linear system responses dependent on the uncertain parameters, a high refinement of the intervals is necessary.

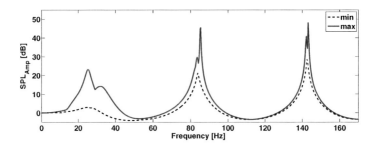

Figure 6.3: Envelopes out of interval endpoints (2^3 samples).

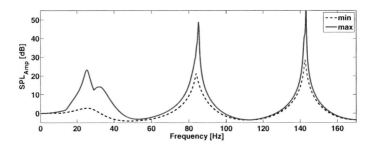

Figure 6.4: Envelopes out of five equal distributed values per interval (5^3 samples).

In addition the number of samples grows exponentially fast, if there are more uncertain parameters. To handle such situations within the interval of uncertainty for the parameters, we have to focus on advanced interval methods, which decrease the needed number of samples [49, 74, 80].

6.2.2 Probabilistic Approach

Probabilistic methods are based on detailed knowledge of the statistical properties of the uncertain variables. The variation in the variable y is specified by the probability density function (pdf) with a related mean value y_0 and a corresponding standard deviation σ_y a priori. Similarly, the variation in the response can be quantified in terms of distribution functions or statistics [49, 73].

For the following example we take the same configuration as in Section 6.2.1. In contrast to the former unknown distribution, all statistical properties of the impedance parameters are known and assumed to be uniformly distributed $U(\cdot, \cdot)$. They are given by

$$
\begin{aligned}
Z(\omega) &= i\left(\omega M - \frac{K}{\omega}\right) + C, & M &\sim U(5\,kg, 15\,kg), \\
K &\sim U(10^5\,\tfrac{N}{m}, 4\cdot 10^5\,\tfrac{N}{m}), & C &\sim U(600\,\tfrac{Ns}{m}, 800\,\tfrac{Ns}{m}).
\end{aligned}
\tag{6.3}
$$

The related mean values and standard deviations follow as

$$
\begin{aligned}
M_0 &= 10\,kg, & \sigma_M &= 2.89\,kg, \\
K_0 &= 2.50 \cdot 10^5\,\tfrac{N}{m}, & \sigma_K &= 8.66 \cdot 10^4\,\tfrac{N}{m}, \\
C_0 &= 700\,\tfrac{Ns}{m}, & \sigma_C &= 57.73\,\tfrac{Ns}{m}.
\end{aligned}
\tag{6.4}
$$

In Figure 6.5 100 frfs out of random combinations of the uncertain impedance parameters, evaluated with the MC method, are presented. The light blue curves describe the frfs each for one sample and the red curve is the resulting mean frf.

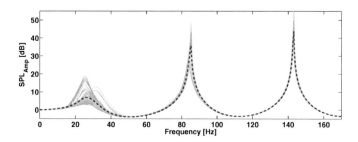

Figure 6.5: 100 frfs (light blue curves) out of MC samples (see equation (6.3)) and the related mean frf (red curve).

In Figure 6.6 the related standard deviation confirms the influence of the parameter uncertainties, especially around the resonances. It is important to notice that the standard

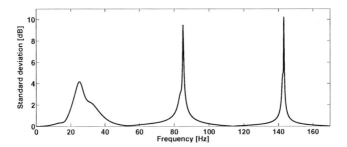

Figure 6.6: Standard deviation of the MC sampled frfs in Figure 6.5.

deviation is huge due to the lengths of intervals in equation (6.3).

In addition to the previous academic example, we consider the single layer absorber made of polyurethane foam, applied in the numerical example in Section 5.1. For this plate-like absorber we assume a realistic uncertainty in the thickness parameter d of the layer, which is uniformly distributed

$$d \sim U(9.5\,cm, 10.5\,cm)\,. \tag{6.5}$$

The corresponding mean value and standard deviation of the thickness are

$$d_0 \;=\; 10\,cm, \qquad \sigma_d \;=\; 0.29\,cm\,. \tag{6.6}$$

In Figure 6.7 we present the frfs and the mean frf out of 100 MC samples. Because of the more realistic interval of uncertainty, the differences in the frfs are smaller in comparison to the previous simulation (Figure 6.5).

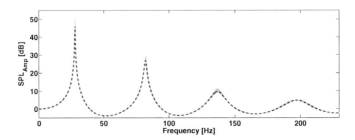

Figure 6.7: 100 frfs (light blue curves) out of MC samples (see equation (6.5)) and the related mean frf (red curve).

The corresponding standard deviation is shown in Figure 6.8. This standard deviation has a more dynamic behaviour in comparison to the previous simulation in Figure 6.6.

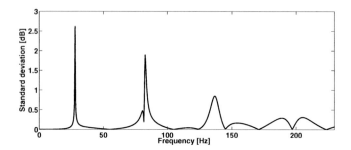

Figure 6.8: Standard deviation of the MC sampled frequency responses in Figure 6.7.

We can explain these non-linear effects partly with the absorption efficiency $\alpha \in [0,1]$ of the single layer polyurethane absorber. The absorption efficiency describes how much percent acoustic energy can be absorbed out of the incident sound waves. Details of absorption efficiency related to compound absorber and absorber parameters can be found in [13]. In Figure 6.9 we present the absorption efficiency in relation to the frequency due to the MC samples in Figure 6.7.

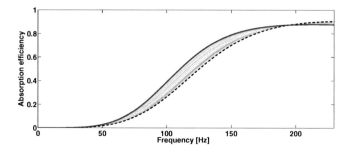

Figure 6.9: Absorption efficiency of the MC thickness samples from equation (6.5).

In Figure 6.9 the blue dotted curve belongs to the minimal thickness sample and the green curve to the maximal thickness sample. The light blue curves belong to the thickness samples between the extremal ones. We observe that in the range from 0 to $195\,Hz$ a thinner absorber is less efficient in absorbing sound waves than a thicker one. From $195\,Hz$ on, this relation becomes inverse and because of this, we notice no deviation at $195\,Hz$ in Figure 6.8. Next to these very basic methods for the possibilistic and the probabilistic approach, there also exist hybrid methods, for example in [61], which seem to be reasonable and compromising for these kind of linear room acoustical problems.

7 Concluding Remarks

At the beginning of the thesis a general overview and a motivation is given to introduce the reader into linear room acoustic problems with different boundary conditions. Then we derived the governing equations from the laws of conservation and from the Euler gas equations to result in the linear Euler gas equations. Subsequently, we presented the governing equations and the different boundary conditions in the state formulation and in the frequency formulation.

In Chapter 3 we identified the subdivided acoustic system as an acoustic network of air, different boundary and source components. In the frame of model order reduction methods the Component Mode Synthesis (CMS) has been applied to approximate the sound field of the interior acoustic system with normal and coupling modes. These normal and coupling modes for the pressure as well as for the velocity are computed out of the velocity potential normal and coupling modes and are the basis functions for the developed coupling method in the frequency domain. Therein normal modes are the eigenmodes of an air component with totally reflecting boundaries and coupling modes enable to couple the interior sound field to absorptive boundary conditions and to adjacent air components. Besides the normal and coupling modes the coupling method is based on Hamilton's Principle to compute the coefficients of the modes by enforcing minimal energy of the enclosed sound field.

To compute the basis functions, the normal and coupling modes, for the coupling method, we proposed numerical methods in Chapter 4. Next to the Spectral Method (SM) the Spectral Finite Element Method (SEM) seems to be the method of choice to compute the velocity potential normal and the velocity potential coupling modes, because it is applicable for arbitrary volumes, yields an exponential convergence rate and is computational efficient due to the related accuracy. The velocity normal and coupling modes and the pressure coupling modes are dependent on derivatives of the velocity potential modes. Because of the C^0 SEM basis expansion it is not possible to gain smooth derivatives out of the velocity potential modes resulting from the SEM. For this problem we applied an adapted patch recovery technique, which takes advantage of the given quadrature points and prescribed boundary values. We explained this method in detail and we worked out its convergence properties.

In Chapter 5 a selection of simulation examples has shown the diversity of applications and the potential of the coupling method.

Finally, we gave a general overview of the uncertainty types, the different uncertainty representations and the methods to propagate uncertainty through numerical methods in Chapter 6. After this general overview we listed the different sources of uncertainty in the considered room acoustical problems and their related effect on the coupling method. In the following we applied the possibilistic and the probabilistic approach to uncertainty in input data to analyse the uncertainty and statistics in the output data.

The main contributions of this work to ongoing research are the development of the coupling method in the frequency domain in Chapter 3 and the chosen and adapted numerical methods presented in Chapter 4. Therein the specific achievements of this thesis are:

- An acoustic network presents the subdivided acoustic system built-up by air components with reflective boundaries, absorber boundaries and transmitting boundaries due to adjacent air components. Inside the air components sound sources are located to force an interior sound field by time harmonic excitation.

- A detailed description of the developed coupling method in the frequency domain has been given. Thereby the interior sound field is approximated by normal and coupling modes for the variables pressure and velocity, which are computed out of velocity potential modes.

- Further more the coupling method is based on Hamilton's Principle to compute the solution by enforcing minimal energy inside the acoustic system.

- The SEM is the choice out of FE methods to compute the steps of the coupling methods algorithm. This specific hp-FE method has been described and implemented to compute the potential velocity modes for arbitrary geometries in one, two and three spatial dimensions.

- In addition, the adapted patch recovery technique enables to compute the pressure coupling modes and the velocity normal and coupling modes out of the derivatives of the velocity potential modes, which is a challenge for general FE solutions in two and three dimensional domains.

- A selection of numerical simulations presents typical applications of the coupling method and its potential for complex problems in room acoustics in the low-frequency range.

- At last the insight into uncertainty analysis and related basic simulations remark that the treatment and propagation of uncertainty is an important issue to gain more realistic results of room acoustic systems.

Mathematical Tools and the Fourier Transform

A.1 Preliminaries

In this section we give some basic definitions and theorems in functional analysis and operator theory [109]. We use these basics to have a mathematical frame for the acoustic systems.

Definition A.1. State variables
The state or primitive variables change in time and describe the state of the underlying system. The variables are scalar quantities $v \in \mathbb{R}$ or vector quantities $\boldsymbol{v} \in \mathbb{R}^3$.

In acoustic systems state variables are the scalar quantities pressure p, mass m, mean sound power W and temperature T. Vector quantities are the sound velocity $\boldsymbol{v}$ and the intensity $\boldsymbol{I}$.

Definition A.2. Conservative variables
The conserved variables of a system are constant in time and are built up from state variables.

Conservative variables in acoustic systems are the scalar quantities energy e and density ρ. The vector quantity is the momentum density $\rho\boldsymbol{v}$, for instance.

Definition A.3. Conservative form
The conservative form of a time dependent system described by partial differential equations is defined as

$$\frac{\partial \boldsymbol{u}}{\partial t} + \nabla^T \cdot (\boldsymbol{F}(\boldsymbol{u})) = \boldsymbol{s}, \quad \boldsymbol{u}, \boldsymbol{s} \in \mathbb{R}^n, \ \boldsymbol{F} : \mathbb{R}^n \to \mathbb{R}^n$$

where $\boldsymbol{u}$ is the vector of conservative variables, $\boldsymbol{F}$ is the flux function related to the conserved variables and $\boldsymbol{s}$ presents external sources, each with respective dimensions.

Definition A.4. Scalar product
Given a K-vector space X, there exists a map $(\cdot, \cdot) : X \times X \to K$ called a scalar product or inner product, if

$$
\begin{aligned}
&i) \quad (\alpha x_1 + \beta x_2, y) = \alpha(x_1, y) + \beta(x_2, y) \quad &&\forall x_1, x_2 \in X, \ \alpha, \beta \in K, \\
&ii) \quad (x, y) = \overline{(y, x)} &&\forall x, y \in X, \\
&iii) \quad (x, x) \geq 0 &&\forall x \in X \ \text{and} \\
& \quad (x, x) = 0 \quad \text{if and only if} \quad x = 0.
\end{aligned}
$$

The scalar product $(\cdot, \cdot)$ is positive definite. In the following we will only consider $K = \mathbb{R}$ or $K = \mathbb{C}$. The scalar product is linear for $K = \mathbb{R}$ and sesquilinear for $K = \mathbb{C}$, i.e. linear in one argument and antilinear in the other one.

Definition A.5. Hilbert space
The Hilbert space H is a real or complex inner product space that is also a complete metric space with respect to the distance function induced by the scalar product. The norm defined by the inner product $(\cdot, \cdot)$ is the real-valued function

$$\|x\| = \sqrt{(x, x)}.$$

The distance between two points $x, y \in H$ is defined by the norm

$$d(x, y) = \|x - y\| = \sqrt{(x - y, x - y)}.$$

In addition the Hilbert space H is complete, if any Cauchy sequence converges with respect to this norm to an element in space.
Let L^2 be the space of square-integrable functions and $V \subset X$, then for every integer $m > 0$, we define the subspace

$$H^m(V) := \left\{ f \in L^2(V) : \partial^i f \in L^2(V), \ i = 0, 1, ..., m \right\},$$

where $\partial^i f$ are the weak derivatives of the function f. A scalar product on $H^m(V)$ is defined by

$$(f, g)_m = \sum_{i=0}^{m} \int_V \partial^i f(x) \overline{\partial^i g(x)} \, dx,$$

inducing the norm $\|f\|_m = (f, f)_m^{\frac{1}{2}}$. The subspaces $H^m(V) \subset L^2(V)$ are again Hilbert spaces.

Definition A.6.
Orthogonality between two distinct elements x, y in a Hilbert space X is given, if $(x, y) = 0$. Two subsets A and B are orthogonal to each other, if every element of A is orthogonal to every element of B. The orthogonal complement of a subset A of X is defined by $A^\perp := y \in X : (x, y) = 0 \ \forall x \in A$. S being a subset of X is a complete orthonormal basis of X, if and only if all elements in S are orthogonal to each other and every element $x \in S$ fulfills $\|x\| = 1$. With $S \subset T$, where T is a orthonormal basis of X as well, follows $S = T$.

Definition A.7. Linear operator, bounded operator, null space, range
A map $A : X \to Y$ is called a linear operator if

$$A(\alpha u + \beta v) = \alpha A(u) + \beta A(v), \quad u, v \in X, \; \alpha, \beta \in \mathbb{C}.$$

The linear operators $V_1 \to V_2$ form the linear space $\mathcal{L}(V_1, V_2)$. The operator A is bounded if there exists a real constant C such that

$$\|A(u)\|_{V_2} = C\|u\|_{V_1}.$$

The bounded operators form a normed linear space $\mathcal{B}(V_1, V_2) \subset \mathcal{L}(V_1, V_2)$. The domain of an operator A is denoted by $D(A)$. The null space $N(A)$ of the operator A is the set

$$N(A) = \{p \in X; \; A(p) = 0\}.$$

The range $R(A)$ of the operator A is defined as

$$R(A) = \{u \in Y; \; \text{there exists } p \in X \text{ such that } A(p) = u\}.$$

Theorem A.1.
Let V and W be Hilbert spaces and $A : D(A) \subset V \to W$ a bounded linear operator. Then $R(A) = W$ if and only if both $R(A)$ is closed and $R(A)^\perp = \{0\}$.

The proof can be found in standard functional-analytical textbooks, for example in [100, 109].

Definition A.8. Closed operator
An operator $A : D(A) \subset V \to W$, where V and W are Banach spaces, is closed if for any sequence $\{v_n\}_{n=1}^\infty \subset D(A)$, $v_n \to v$ and $A(v_n) \to w$ imply that $v \in D(A)$ and $w = Av$.

Theorem A.2.
Let V and W be Hilbert spaces and $A : D(A) \subset V \to W$ be a bounded linear operator. Assume that there exists a constant $C > 0$ such that

$$\|Ap\|_W \geq C\|v\|_V \qquad \forall v \in D(A).$$

This inequality sometimes is called the stability or coercivity estimate. If $R(A)^\perp = \{0\}$ then the operator equation $Au = f$ has a unique solution.

Proof:
Let us verify that $R(A)$ is closed. Let $\{w_n\}_{n=1}^\infty \subset R(A)$ with $w_n \to w$. Then there is a sequence $\{v_n\}_{n=1}^\infty \subset D(A)$ such that $w_n = Av_n$. The previous estimate implies $C\|v_n - v_m\|_V \leq \|w_n - w_m\|_W$, which is equivalent to that $\{v_n\}_{n=1}^\infty$ is a Cauchy sequence in V. Completeness of the Hilbert space V yields existence of a $v \in V$ such that $v_n \to v$. Since A is closed, we obtain $v \in D(A)$ and $w = Av \in R(A)$. The previous theorem yields the existence of a solution and the uniqueness of the solution follows immediately from the previous stability estimate.

A.2 Fourier Transform

The Fourier transform and its generalized variants have an enormous practical meaning in many fields of science and industry [8]. In acoustics for example the Fourier transform is the frequency transform to decompose the time-dependent sound into harmonics.

A.2.1 Continuous Fourier Transform

The continuous Fourier transform (FT) is an operation, that transforms one complex-valued function of a real variable ($g : \mathbb{R}^n \to \mathbb{C}$) into another complex-valued function ($\hat{g} : \mathbb{R}^n \to \mathbb{C}$). g is the representation for the time variable in the time domain and $\hat{g}$ is the frequency domain representation of the original function g. $\hat{g}$ describes which frequencies are present in the original function. The term FT refers both to the frequency domain representation of a function, and to the process or formula that transforms one function g into the other $\hat{g}$. In this specific case, the continuous FT, both the time and frequency domains are unbounded linear continua.

Definition A.9. Fourier transform
The FT of an integrable function $g \in L^1$ with $x : \mathbb{R} \to \mathbb{C}$ is defined as:

$$\hat{g}(f) = \int\limits_{-\infty}^{\infty} g(t)e^{-i2\pi ft}\, dt, \qquad \forall \xi \in \mathbb{R}\,.$$

Then g can be reconstructed with $\hat{g} : \mathbb{R} \to \mathbb{C}$ by the inverse Fourier transform (IFT) with

$$g(t) = \int\limits_{-\infty}^{\infty} \hat{g}(f)e^{i2\pi ft}\, df, \qquad \forall t \in \mathbb{R}\,.$$

The FT is linear, invertible, continuous, bounded and converges for $|f| \to \infty$ to zero. The existence as well as the boundedness of $\hat{g}$ follows from

$$|\hat{g}(f)| \leq = \int\limits_{-\infty}^{\infty} |g(t)e^{-i2\pi ft}|\, dt \leq \int\limits_{-\infty}^{\infty} |g(t)|\, dt\,. \tag{A.1}$$

When the independent variable t represents time (in seconds), the transformed variable ξ represents the frequency.
There are several common conventions for defining the FT. Another common notation for the FT is the following

$$\hat{g}(\boldsymbol{\omega}) = \tfrac{1}{(2\pi)^{\frac{n}{2}}} \int\limits_{\mathbb{R}^n} g(\boldsymbol{x})e^{i\boldsymbol{\omega}^T \boldsymbol{x}} dx, \qquad \forall \boldsymbol{x} \in \mathbb{R}^n$$

$$f(\boldsymbol{x}) = \tfrac{1}{(2\pi)^{\frac{n}{2}}} \int\limits_{\mathbb{R}^n} \hat{f}(\boldsymbol{\omega})e^{-i\boldsymbol{\omega}^T \boldsymbol{x}} d\boldsymbol{\omega}, \qquad \forall \boldsymbol{\omega} \in \mathbb{R}^n\,.$$

In one dimension $(n = 1)$ ω stands for the circular frequency with $\omega = 2\pi f$ with units rad per second.

The motivation of the FT comes from the study of Fourier series. We suppose, that f is a function, which is zero outside of some interval $[-\frac{L}{2}, \frac{L}{2}]$. Then for any $T \geq L$ we expand f in a Fourier series on the interval $[-\frac{T}{2}, \frac{T}{2}]$:

$$\hat{g}(\tfrac{n}{T}) = \int\limits_{-\frac{T}{2}}^{\frac{T}{2}} g(t)e^{-2\pi i \frac{n}{T}x}\,dt, \qquad \forall n \in \mathbb{Z}$$

$$g(t) = \frac{1}{T}\sum\limits_{n=-\infty}^{\infty} \hat{g}(\tfrac{n}{T})e^{2\pi i \frac{n}{T}t}, \qquad \forall n \in \mathbb{Z}.$$

If we let $f_n = \frac{n}{T}$ and $\Delta f = \frac{(n+1)-n}{T} = \frac{1}{T}$, then the last sum becomes the Riemann sum

$$g(t) = \sum\limits_{n=-\infty}^{\infty} \hat{g}(f_n)e^{i2\pi f_n t}\Delta f, \qquad \forall n \in \mathbb{Z}.$$

For $T \to \infty$ this Riemann sum converges to the integral for the inverse FT.

A.2.2 Discrete-Time Fourier Transform

The discrete-time Fourier transform (DTFT) is a specific form of the FT [12]. In contrast to the FT, the DTFT requires an input function that is continuous and periodic related to non-zero values on a finite time duration. Consequently one period of the function contains all of the unique information. The DTFT needs sampled discrete values of the continuous function g and the resulting frequency domain representation $\hat{g}$ has a discrete spectrum and a finite length.

Definition A.10. Discrete-time Fourier transform
Given a discrete set of sampled real or complex function values $g_n = g(nT)$, $n \in \mathbb{Z}$ of a continuous function g the DTFT transformed of g is written as

$$\hat{g}(\omega) = \sum\limits_{n=-\infty}^{\infty} g_n e^{-i\omega n} \quad or \quad \hat{g}(f)_T = T\sum\limits_{n=-\infty}^{\infty} g(nT)e^{-i2\pi fTn}\,.$$

Here g_n denote the function values for discrete moments in time $f_n = f(nT)$, where T is the sampling interval. $\omega = 2\pi fT$ and $\frac{1}{T} = f_s$ is the sampling rate.
The inverse time-discrete Fourier transform (IDTFT) reads as follows

$$g_n = \frac{1}{2\pi}\int\limits_{-\pi}^{\pi} \hat{f}(\omega)e^{i\omega n}\,d\omega \quad or \quad g(nT) = T\int\limits_{-\frac{1}{2T}}^{\frac{1}{2T}} \hat{f}_T(f)e^{i2\pi fnT}\,df\,.$$

113

f represents the ordinary frequency in cycles per second and f_s, the sampling rate, has units of samples per second. In $\omega = 2\pi f T = 2\pi \frac{f}{f_s}$ the units of $\frac{f}{f_s}$ are cycles per sample. It is common to replace this ratio with a single variable, called normalized frequency, which represents actual frequencies as multiples of the sample rate. Consequently ω is also a normalized frequency, but its units are radians per sample. The normalized frequency has the advantage that the function $\hat{g}(\omega)$ is periodic with a period 2ω. Thats the reason why the inverse transform needs only be evaluated in the interval 2π. The alternate notation $\hat{g}(e^{i\omega})$ for the DTFT $\hat{g}(\omega)$ emphasizes the periodicity property and helps us to distinguish between the DTFT and underlying Fourier transform of $g(t)$, and highlights the relation of the DTFT to the z-transform [8].

A.2.3 Discrete Fourier Transform

The discrete Fourier transform (DFT) is a specific kind of the DTFT. In contrast to the DTFT, the DFT requires an periodic input function that is discrete and whose non-zero values have a limited (finite) duration. Using the DFT implies, that the finite analysed segment is one period of an infinitely extended periodic signal. If that is not the case, we get artifacts at the boundary, the so-called Gibb's phenomenon [39, 40]. For that we have to use a suited window function to reduce the artifacts in the resulting frequency domain representation.

Definition A.11. Discrete Fourier transform
The sequence of N complex function values $g_n = g(t_n)$ with $g_n = g_{n+N}$ at the points $t_n = n\Delta t = n\frac{L}{N}$ with $-\frac{L}{2} \leq t \leq \frac{L}{2}$ is transformed into the sequence of N complex values $\hat{g}_k$ by the DFT due to the formula:

$$\hat{g}_k = \sum_{n=0}^{N} g_n e^{-i2\pi k \frac{n}{N}}, \qquad k = 0, ..., N-1.$$

The inverse discrete Fourier transform (IDFT) denotes

$$g_n = \frac{1}{N} \sum_{k=0}^{N} \hat{g}_k e^{i2\pi k \frac{n}{N}}, \qquad n = 0, ..., N-1.$$

We can interpret the complex numbers $\hat{g}_k$ as representatives of the amplitude and phase of the different sinusoidal components of the input signal g_n.

We can also write this transform in terms of a DFT matrix $\boldsymbol{W}$ with

$$\boldsymbol{W} = \begin{pmatrix} \omega_N^{0 \cdot 0} & \cdots & \omega_N^{0 \cdot (N-1)} \\ \vdots & \ddots & \vdots \\ \omega_N^{(N-1) \cdot 0} & \cdots & \omega_N^{(N-1) \cdot (N-1)} \end{pmatrix}$$

with $\omega_N = e^{-i2\pi\frac{1}{N}}$. With the input values $\boldsymbol{g} = (g_0, \ldots g_{N-1})^T$ and the output values $\hat{g} = (\hat{g}_0, \ldots \hat{g}_{N-1})^T$ we arrive at the DFT matrix vector equation

$$\hat{g} = \boldsymbol{W}\boldsymbol{g}. \tag{A.2}$$

When the DFT matrix $\boldsymbol{W}$ is scaled with $\frac{1}{\sqrt{N}}$, it becomes a unitary matrix, which is due to the orthogonality relation

$$\sum_{n=0}^{N_1} \left(e^{i2\pi k\frac{n}{N}}\right) \left(e^{-i2\pi k'\frac{n}{N}}\right) = N\delta_{kk'}. \tag{A.3}$$

That means the vectors $e^{2\pi i k\frac{n}{N}}$ form an orthogonal basis over the set of N-dimensional complex vectors. A key enabling factor is the fact that the DFT can be computed efficiently in practice using a fast Fourier transform (FFT) algorithm [21, 107].

B Impedance Boundary Condition

Next to the general mathematical description of boundary conditions in acoustics, we want to give the reader a more detailed insight into the boundary condition of interest, the impedance boundary condition. This is the most realistic one and is of special importance for the underlying work.

The dynamic behaviour of a system is evaluated partly by the characteristics of the individual components and partly by the dynamic interactions between them. If the behaviour of assemblages of different components is of interest, then the concept of impedance and its complementary quantity the mobility comes into play. The system formed by components and junctions between them can be presented in an acoustic network, what we will discuss in the following chapter in more detail. Interactions between components involve the incidence of vibrational or acoustic waves upon the junctions, connections and interfaces between components, together with their reflection and transmission. Thereby, wave energy dissipates partly into heat. Impedance has a great practical importance in controlling the behaviour of vibroacoustic systems such as increasing the efficiency of generated sound energy or impede wave transmission amongst many others [32, 37].

Although this is not the general case, the Fourier transformation allows to apply the concept of impedance in all linear cases. The imaginary part of the impedance is referred to as the reactance and the real part as the resistance. As we shall see, impedance comes in many guises [33], which will be presented in the following.

B.1 Mechanical Impedance

Mechanical impedance related to solid structures

The viscously damped single-degree-of-freedom system in Figure B.1 combines a lumped mass, a massless spring, and a damper element. Here the mechanical point impedance Z is defined as follows

$$Z(\omega) = i\left(\omega M - \frac{K}{\omega}\right) + C, \quad \{M, K, C\} \in \mathbb{R}_0^+. \tag{B.1}$$

In equation (B.1) M, K, and C denote the mass of the mass element, the stiffness of the spring element, and the damping coefficient of the damper element of the system, respectively (see Figure B.1). ω again stands for the circular frequency of excitation by a harmonic force acting on the system. To evaluate the natural frequency of the system, we determine the following expression $max \, |i\omega Z(\omega)|^{-2}$.

The real part of a point impedance is always positive, because it is related to dissipation. The imaginary part can be positive (mass-like) or negative (spring-like). The point impedance (B.1) is stiffness-dominated at low frequencies, damping-dominated at resonances, and mass-dominated at high frequencies.

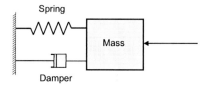

Figure B.1: Ideal viscous damper oscillator.

Mechanical radiation impedance

The mechanical radiation impedance (Figure B.2) of an oscillating rigid structure is represented as

$$Z = \frac{\int_S p \, dS}{v} = \frac{F}{v}, \tag{B.2}$$

where v is the velocity of the rigid body and F is the force applied to the fluid.

An infinite plate impedance, for example, is real and independent of frequency (equivalent to a damper). A beam impedance has a damper part and a mass part, both frequency dependent. The impedance of a finite structure tends to that of the equivalent infinite structure at high frequency.

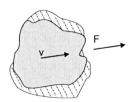

Figure B.2: Mechanical radiation impedance of an oscillating rigid body.

B.2 Acoustic Impedance

Acoustic impedance related to fluids

The acoustic impedance Z stands for the ratio of complex amplitude of harmonic pressure, or force on a surface, to the associated particle velocity, or associated volume velocity through a surface. Furthermore, the acoustic impedance determines the degree to which sound waves in one region (of fluid) are reflected and transmitted at the interface to another fluid. The most important contribution of the underlaying work is the analysis of the interaction between fluid and solid systems related to sound absorption, reflection, and transmission. In addition further focus is laid on the fluid loading imposed upon vibrating solids. It is worth mentioning that the terminology varies from book to book and causes a common confusion.

The characteristic acoustic impedance $Z_0 = \rho_0 c_0$ is a material property and depends on the density ρ and the longitudinal wave speed c. For example, with a given reference temperature of $20°C$, a reference density $1.2 \frac{kg}{m^3}$, and the related reference speed of sound $343.4 \frac{m}{s}$ we obtain the characteristic impedance $Z_0 = 413.5 \frac{Ns}{m^3}$.

Normal specific acoustic impedance

In the remainder of this work, we will focus on the (local reacting) normal specific acoustic impedance of a boundary, which is defined as

$$Z = \frac{p}{\boldsymbol{v}^T \boldsymbol{n}} \,. \tag{B.3}$$

For the purpose of expressing the interaction between an incident sound wave and the material, it is necessary to specify the normal specific acoustic impedance of the surface. This normal specific acoustic impedance is defined as the ratio of the complex amplitude of surface pressure to that of the component of particle velocity normal to, and directed into, the surface for harmonically varying quantities. This form of impedance is also known as boundary impedance and is represented exemplary in Figure B.3.

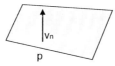

Figure B.3: Normal specific acoustic impedance of a boundary.

Acoustic radiation impedance

In analogy to (B.2), it is worth defining the acoustic radiation impedance when neither the normal particle velocity nor the pressure are uniform over some fluid surface

$$Z = \frac{\left(\frac{1}{S}\right) \int_S p \, dS}{\int_S \boldsymbol{v}^T \boldsymbol{n} \, dS} \,. \tag{B.4}$$

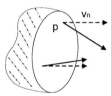

Figure B.4: Acoustic impedance of a small surface acting in a fluid.

Bibliography

[1] A. Ahmadi, K.S. Surana, R.K. Maduri, A. Romkes, and J.N. Reddy. Higher order global differentiability local approximations for 2-d distorted quadilateral elements. *International Jornal for Computational Methods in Engineering Science and Mechanics*, 10:1–19, 2009.

[2] B.M. Ayyub and G.J. Klir. *Uncertainty Modeling and Analysis in Engineering and the Sciences*. Crc Pr Inc, 2006.

[3] I. Babuska. The Finite Element Method With Lagrangian Multipliers. *Numerische Mathematik*, 20:179–192, 1973.

[4] I. Babuska and M. Suri. The p and h-p versions of the finite element method, basic principles and properties. *SIAM Review*, 36:578, 1994.

[5] M.T. Bah, P.B. Nair, A. Bhaskar, and A.J. Keane. Stochastic component modes synthesis. In *44th AIAA/ASME/ASCE/AHS/ASC Structures, Structural Dynamics, and Materials Conference, Norfolk, Virginia, Apr. 7-10*, volume 1750. AIAA, 2003.

[6] H. Bandemer, editor. *Modelling Uncertain Data*. Wiley-VCH Verlag GmbH, 1993.

[7] K.-J. Bathe and H. Zhang. Finite element developments for general fluid flows with structural interactions. *International Journal for Numerical Methods in Engineering*, 60:213–232, 2004.

[8] R.J. Beerends, H.G. Morsche, and J.C. van den Berg. *Fourier and Laplace Transforms*. Cambridge University Press, 2003.

[9] S.C. Brenner and L.R. Scott. *The Mathematical Theory of Finite Element Methods*. Springer, 1996.

[10] F. Brezzi. On the existence, uniqueness and approximation of saddles point problems arising from lagrangian multipliers. *RAIRO Analyse Numérique*, 8:129–151, 1974.

[11] F. Brezzi and M. Fortin. *Mixed and Hybrid Finite Element Methods*. Springer, Berlin, 1991.

[12] W.L. Briggs and V.E. Henson. *The DFT - The Owner's Manual for the Discrete Fourier Transform*. SIAM, 1995.

[13] M. Buchschmid. *ITM-based FSI-models for rooms with absorptive boundaries (preprint)*. PhD thesis, Technische Universität München, 2011.

[14] M. Buchschmid and G. Müller. Modeling of wave number dependent absorptive characteristics with the help of the theory of porous media. In *EURODYN Conference*, 2008.

[15] M. Buchschmid, M. Pospiech, and G. Müller. A semi - analytical model for rooms with absorptive boundary conditions. In *DAGA*, 2009.

[16] M. Buchschmid, M. Pospiech, and G. Müller. ITM-based FSI-models for applications in room acoustics. In *COMPDYN Conference*, 2009.

[17] M. Buchschmid, M. Pospiech, and G. Müller. Coupling Impedance Boundary Conditions for Absorptive Structures with Spectral Finite Elements in Room Acoustical Simulations. *Journal of Computing and Visualization in Science - Special Issue: Hot Topics in Computational Engineering, Springer*, 2010 (preprint).

[18] H.-J. Bungartz and M. Schäfer, editors. *Fluid-structure interaction: Modelling, Simulation, Optimization*. Springer, 2006.

[19] C. Canuto, M.Y. Hussaini, and A. Quarteroni. *Spectral Methods in Fluid Dynamics*. Springer, Berlin, 1988.

[20] P.G. Ciarlet. *The Finite Element Method for Elliptic Problems*. North-Holland, Amsterdam, 1978.

[21] J. Cooley and J. Tukey. An algorithm for the machine calculation of complex fourier series. *Mathematics of Computation*, 19:297–301, 1965.

[22] R. Courant and D. Hilbert. *Methods of Mathematical Physics*, volume 1. Wiley: New York, 1989.

[23] R.R.J. Craig. Methods of component mode synthesis. *Shock and Vibration Digest*, 9:3–10, 1977.

[24] R.R.J. Craig. A review of time-domain and frequency-domain component mode synthesis methods. *Journal of Modal Analysis*, 2:59–72, 1987.

[25] R.R.J. Craig and M.C.C. Bampton. Coupling of substructures for dynamic analyses. *AIAA*, 6:1313–1319, 1968.

[26] R.R.J. Craig and C.J. Chang. Free-interface methods of substructure coupling for dynamic analysis. *AIAA Journal of Computational Physics*, 14:1633–1635, 1976.

[27] L. Cremer, M. Heckl, and B.A.T. Petersson. *Structure-Borne Sound: Structural Vibrations and Sound Radiation at Audio Frequencies*. Springer, 3rd edition, 2005.

[28] L. Cremer and H. Müller. *Die wissenschaftlichen Grundlagen der Raumakustik, Band II: Wellentheoretische Raumakustik*. Hirzel, Stuttgart, 1976.

[29] R. de Boer. *Theory of Porous Media. Highlights in Historical Development and Current State*. Springer, 2000.

[30] R. de Boer. *Trends in Continuum Mechanics of Porous Media. Theory and Applications of Transport in Porous Media.* Springer, 2005.

[31] W.M. Dong, W.L. Chiang, and F.S. Wong. Propagation of uncertainties in deterministic systems. *Computers and Structures*, 25(3):415–423, 1987.

[32] F.A. Everest and K.C. Pohlmann. *Master Handbook of Acoustics.* McGrawHill, 5. edition, 2009.

[33] F. Fahy. *Foundations of Engineering Acoustics.* Elsevier Academic Press, 2007.

[34] F. Fahy and P. Gardonio. *Sound and Structural Vibration - Radiation, Transmission and Response.* Elsevier Academic Press, 2nd edition, 2007.

[35] H. Fischer. The central limit theorem from laplace to cauchy: Changes in stochastic objectives and in analytical methods.

[36] G.S. Fishman. *Monte Carlo: Concepts, algorithms, and applications.* Springer Series in Operations Research, Springer-Verlag, Berlin, 1996.

[37] H.V. Fuchs. *Schallabsorber und Schalldämpfer.* Springer, 3. edition, 2010.

[38] R.G. Ghanem and P.D. Spanos. *Stochastic Finite Elements: A Spectral Approach.* Dover Publications, 2003.

[39] J.W. Gibbs. Fourier's series. *Nature*, 59:200–200, 1898.

[40] J.W. Gibbs. Fourier's series. *Nature*, 59:606–606, 1899.

[41] D. Givoli. On the number of reliable finite-elemtent eigenmodes. *Communications in Numerical Methods in Engineering*, 24:1967–1977, 2008.

[42] D. Gottlieb and S.A. Orszag. *Numerical Analysis of Spectral Methods: Theory and Applications.* SIAM, Philadelphia, 1977.

[43] M. Grigoriu. *Stochastic Calculus: Application in Science and Engineering.* Birkhäuser, 2002.

[44] A. Haldar and S. Mahadevan. *Reliability Assessment Using Stochastic Finite Element Analysis.* Wiley & Sons, 2000.

[45] I. Harari. A survey of finite element methods for time-harmonic acoustics. *Computer Methods in Applied Mechanical Engineering*, 195:1594–1607, 2006.

[46] I. Harari and T.J.R. Hughes. A cost comparison of boundary element and finite element methods for problems of time-harmonic acoustics. *Computer Methods in Applied Mechanical Engineering*, 97:77–102, 1992.

[47] A. Hepberger, H.H. Priebsch, W. Desmet, B. van Hal, and P. Sas. Application of the wave based method for the steady-state acoustic response prediction of a car cavity in the mid-frequency range. In *ISMA-International Conference on Noise and Vibration Engineering, KUL, Belgium*, 2002.

[48] Ester Hills. *Uncertainty Propagation in Structural Dynamics with Special Reference to Component Modal Models.* PhD thesis, University of Southamptonn, Faculty of Engineering, Science and Mathematics, Institute of Sound and Vibration Research, 2006.

[49] L. Hinke. *Modelling approaches for the low-frequency analysis of built-up structures with non-deterministic properties.* PhD thesis, University of Southampton, Faculty of Engineering, Science and Mathematics, Institute of Sound and Vibration Research, 2008.

[50] M.S. Howe. *Acoustics of fluid-structure interactions.* Cambridge Monographs on Mechanics, 2004.

[51] T.J.R. Hughes. *The Finite Element Method.* Dover Publications: New York, 1987.

[52] T.J.R. Hughes, A. Reali, and G. Sangalli. Duality and unified analysis of discrete approximations in structural dynamics and wave propagation: Comparison of p-method finite elements with k-method nurbs. *Computer Methods in Applied Mechanics and Engineering,* 197:4104–4124, 2008.

[53] T.J.R. Hughes, A. Reali, and G. Sangalli. Isogeometric methods in structural dynamics and wave propagation. In *Compdyn 2009,* 2009.

[54] F. Ihlenburg. *Finite Element Analysis of Acoustic Scattering.* Applied Mathematical Sciences 132. Springer, 1998.

[55] D.C. Simkins Jr., S. Li, H. Lu, and W.K. Liu. Reproducing kernel element method Part IV: Globally compatible C^n ($n \geq 1$) triangular hierarchy. *Computer Methods in Applied Mechanics and Engineering,* 193:1013–1034, 2004.

[56] H. Kadner. Untersuchungen zur Kollokationsmethode, aus: Dissertation v. 17. Februar 1958. *ZAMM-Journal of Applied Mathematics and Mechanics,* 40:99–113, 2006.

[57] G.E. Karniadakis and S. Sherwin. *Spectral/hp Element Methods for computational Fluid Dynamics,* volume 2. Oxford University Press, 2005.

[58] A.J. Keane and W.G. Price. *Statistical Energy Analysis: An Overview with Applications in Structural Dynamics.* Cambridge Universtiy Press, Cambridge, UK, 1997.

[59] J.G. Klir and B. Yuan. *Fuzzy sets and Fuzzy Logic: Theory and Applications.* Prentice Hall, Englewood Cliffs, New Jersey, 1995.

[60] R.S. Langley. Recent advances and remaining challenges in the statistical energy analysis. In *EURODYN,* 2008.

[61] R.S. Langley, R. Lande, P.J. Shorter, and V. Cotoni. Hybrid deterministic-statistical modelling of built-up structures. In *ICSV 12,* 2005.

[62] W. Larbi, J.-F. Deü, and R. Ohayon. Finite element formulations for structural-acoustic internal problems with poroelastic treatment. In *COMPDYN Conference*, 2009.

[63] S. Larsson and V. Thomée. *Partial Differential Equations with Numerical Methods*. Springer, 2008.

[64] R. LeVeque. *Numerical Methods for Conservation Laws*. Birkhäuser Basel, 1992.

[65] S. Li, H. Lu, W. Han, W.K. Liu, and D.C. Simkins. Reproducing kernel element method Part II: Globally conforming I^m/C^n hierarchies. *Computer Methods in Applied Mechanics and Engineering,*, 193:953–987, 2004.

[66] M.J. Lighthill. On sound generated aerodynamically. *Proceedings of the Royal Society of London, Series A*, 211:564–587, 1952.

[67] M.J. Lighthill. *An introduction to Fourier analysis and generalised functions*. Cambridge University Press, 2003.

[68] C.-H. Lin. Improvement of vibration problems using Helmholtz equation to recover nodal gradients. *Engineering Computations*, 15(3):288–311, 1998.

[69] H. Lu, S. Li, D.C. Simkins Jr., and W.K. Liu. Reproducing kernel element method Part III: Generalized enrichment and applications. *Computer Methods in Applied Mechanics and Engineering*, 193:989–1011, 2004.

[70] R.H. Lyon and R.G. DeJong. *Theory and Application of Statistical Energy Analysis*. Butterworth-Heinemann, Boston, 2 edition, 1995.

[71] Y. Maday and A.T. Patera. *Spectral element methods for the Navier-Stokes equations. In State-of-the-art surveys in computational mechanics*. ASME, New York, 1989.

[72] F. Magoulis, editor. *Substructuring Techniques and Domain Decomposition Methods*. Saxe-Coburg Publications, 2010.

[73] O.P. Le Maître and O.M. Knio. *Spectral Methods for Uncertainty Quantification, With Application to Computational Fluid Dynamics*. Springer, 2010.

[74] G. Manson. Calculating frequency response functions for uncertain systems using complex affine analysis. *Journal of Sound and Vibration*, 288:487–521, 2005.

[75] S. Marburg and B. Nolte. *Computational Acoustics of Noise Propagation in Fluids: Finite and Boundary Elements Methods*. Springer, Berlin, 2008.

[76] MATLAB - MATrix LABoratory. http://www.mathworks.com/products/matlab/.

[77] M.D.C.Magalhaes and N.S. Ferguson. The development of a component mode synthesis (cms) model for three-dimensional fluid-structure interaction. *Acoustical Society of America*, 118 (6):3679–3690, 2005.

[78] O.Z. Mehdizadeh and M. Paraschivoiu. Investigation of a two-dimensional spectral element method for helmholtz's equation. *Journal of Computational Physics*, 189:111–129, 2003.

[79] G. Müller and M. Möser. *Handbook of Engineering Acoustics: A Handbook*. Springer, 2010.

[80] D. Moens, H. De Gersem, W. Desmet, and D. Vandepitte. An overview of novel non-probabilistic approaches for non-deterministic dynamic analysis. In *NOVEM*, 2005.

[81] H.J.-P. Morand and R. Ohayon. *Fluid Structure Interaction*. Wiley, 1995.

[82] P.B. Nair and A.J. Keane. Stochastic reduced basis methods. *AIAA*, 40(8):1653–1664, August 2002.

[83] J. T. Oden, S. Prudhomme, and L. Demkowicz. A posteriori error estimation for acoustic wave propagation problems. *Archives of Computational Methods in Engineering*, 12:343–389, 2005.

[84] Technische Universität München Chair of Structural Mechanics. http://www.bm.bv.tum.de/.

[85] R. Ohayon and C. Soize. *Structural acoustics and vibration*. Academic Press, New York, 1998.

[86] S.-A. Papanicolopulos, A. Zervos, and I. Vardoulakis. A three-dimensional C^1 finite element for gradient elasticity. *International Journal for Numerical Methods in Engineering*, 77:1396–1415, 2009.

[87] C. Petersen. *Statik und Stabilität*. vieweg, 1982.

[88] C. Petersen. *Dynamik der Baukonstruktionen*. vieweg, 2000.

[89] L. Piegl and W. Tiller. *The NURBS book*. Springer, 2000.

[90] Y. Pinchover and J. Rubinstein. *An Introduction to Partial Differential Equations*. Cambridge University Press: Cambridge, 2005.

[91] M. Pospiech, M. Buchschmid, and G. Müller. Spectral approaches for room acoustics. In *ICSV 16*, 2009.

[92] C. Pozrikidis. *Finite and Spectral Element Methods using MATLAB*. Chapman & Hall/CRC, 2005.

[93] Z.-Q. Qu. *Model Order Reduction Techniques: With Applications in Finite Element Analysis*. Springer, 2004.

[94] S.W. Rienstra and A. Hirschberg. *An Introduction to Acoustics*. lecture script, 2010.

[95] S.K. Sachdeva. *Subspace Projectin Schemes for Stochastic Finite Element Analysis.* PhD thesis, Department of Mechanical Engineering, University of Southampton, 2006.

[96] W.H. Schilders, H.A. van der Vorst, and J. Rommes, editors. *Model Order Reduction: Theory, Research Aspects and Applications (Mathematics in Industry).* Springer, 2008.

[97] Hans R. Schwarz. *Methode der Finiten Elemente.* Teubner Verlag, 1984.

[98] W.-H. Shyu, Z.-D. Ma, and G.M. Hulbert. A new component mode synthesis method: Quasi-static mode compensation. *Finite Elements in Analysis and Design,* 24:271–281, 1997.

[99] D.C. Simkins, A. Kumar, N. Collier, and L.B. Whitenack. Geometry representation, modification and iterative design using rkem,. *Computer Methods in Applied Mechanics and Engineering,* 196:4304–4320, 2007.

[100] P. Solin. *Partial Differential Equations and the Finite Element Method.* WILEY, 2006.

[101] J. Sremčević. *Model order reduction methods in room acoustics (preprint).* PhD thesis, Technische Universität München, 2011.

[102] J. Stoer and R. Bulirsch. *Numerische Mathematik 2: Eine Einführung - unter Berücksichtigung von Vorlesungen von F.L. Bauer.* Springer, 2005.

[103] G. Strang and G.J. Fix. *An Analysis of the Finite Element Method.* Pentice-Hall: Englewood Cliffs, NJ, 1973.

[104] W.A. Strauss. *Partial Differential Equations.* Wiley, 2nd edition, 2007.

[105] L.L. Thompson. A review of finite-element methods for time-harmonic acoustics. *JASA,* 119:1315–1330, 2006.

[106] L. N. Trefethen. *Spectral Methods in Matlab.* SIAM, 2000.

[107] C. van Loan. *Computational Framework for the Fast Fourier Transform.* SIAM, 1992.

[108] I. Veit. *Technische Akustik.* Vogel, 6. edition, 2005.

[109] D. Werner. *Funktionalanalysis.* Springer, 5. edition, 2005.

[110] O.C. Zienkiewicz. *The Finite Element Method, Vol. 1: The Basis.* Butterworth-Heinemann, Oxford, 5. edition, 2000.